ENERGY AND AMERICAN SOCIETY

A Reference Handbook

ENERGY AND AMERICAN SOCIETY

A Reference Handbook

E. Willard Miller
Department of Geography

Ruby M. Miller
Pattee Library

The Pennsylvania State University

CONTEMPORARY WORLD ISSUES

ABC-CLIO

Santa Barbara, California
Denver, Colorado
Oxford, England

Library of Congress Cataloging-in-Publication Data

Miller, E. Willard (Eugene Willard), 1915–
    Energy and American society : a reference handbook /
  E. Willard Miller, Ruby M. Miller.
      p. cm. — (Contemporary world issues)
    Includes bibliographical references and index.
    1. Power resources—United States—Handbooks, manuals, etc.
  I. Miller, Ruby M. II. Title. III. Series.
  TJ163.235.M55   1993   333.79'0973—dc20   93–30307

ISBN 0-87436-689-5 (alk. paper)

00 99 98 97 96 95 94 93   10 9 8 7 6 5 4 3 2 1 (cloth)

ABC-CLIO, Inc.
130 Cremona Drive, P.O. Box 1911
Santa Barbara, California 93116-1911

This book is printed on acid-free paper ∞ .
Manufactured in the United States of America

Contents

4   Directory of Organizations, 201

5   Bibliography, 245

6 Films and Videocassettes, 353

Preface

Energy is the catalyst that created the Industrial Revolution about 250 years ago. More than any other single factor, energy has shaped our modern culture. The United States in its infancy was magnificently endowed with fossil fuels. For decades the general public believed that our resources were inexhaustible. While the present reserve of coal will last for several hundred more years, petroleum reserves are known to be more limited, and the nation currently imports about half its needs—making all citizens more aware that reserves of many energy resources are indeed limited. In addition, the use of fossil fuels presents many environmental problems. In the early days of the development of nuclear energy, many thought this new source of energy would replace fossil fuels. Because of the inherent dangers of radioactivity, the development of nuclear energy has diminished its role in the American energy scenario. In spite of the need to find alternatives to traditional energy resources, the energy obtained from renewable sources such as the wind, the sun, geothermal activity, and the like is relatively insignificant.

This volume begins with a perspective of energy sources available to the nation. A number of problems in different energy realms are investigated. For example, under the discussion about petroleum, the role of the Organization of Petroleum Exporting Countries (OPEC) and the response of the American government to the oil crisis of the 1970s are analyzed. Although coal played a dominant role in supplying the energy needs of the nineteenth and early twentieth centures, the competition between coal and petroleum has created many economic problems for the coal industry in recent decades. A special section is devoted to the use and decline of the anthracite coal industry. Although the reserves of anthracite remain great, anthracite plays no role in the modern energy economy.

Other chapters look at a variety of topics that amplify the examination of the role of energy in modern American society. The chronologies and statistics in Chapter 2 review the evolution of energy discovery and demand in the United States, including trends in reserves and production, major American petroleum companies, OPEC nations, and evolution of energy laws and regulations as well as significant tabular and statistical data. Chapter 3 describes laws and regulations passed by the federal government in order to aid energy industries, and Chapter 4 provides an organizational directory divided into four major parts: The first lists U.S. government organizations; the second, the U.S. intergovernmantal advisory committees; the third, individual energy organizations; and the fourth, international organizations.

In recent years the amount of literature on energy has increased dramatically. The literature varies from scientific to political and economic. The bibliography section (Chapter 5) includes about 100 annotated book citations and several hundred journal articles and government documents, presenting a wide range of material on energy fields. At the end of the bibliography, there is a list of selected journals that publish articles on energy topics. The volume concludes with an annotated list of films dealing with energy and an author/title/subject index.

E. Willard Miller
Ruby M. Miller

1

Energy: A Perspective

THE UNITED STATES COULD NOT HAVE ACHIEVED its industrial position, nor could the citizens of our country have attained their high standard of living without the development of the nation's rich energy sources. The energy resources provide a base for the nation's military and economic strength. Because of the vast quantity of fossil fuels, there has been little effort to utilize these resources efficiently throughout most of our history. Only in recent years has it become evident that these resources are finite and now must be used with greater efficiency.

Patterns of Energy Evolution

The pattern of fuel consumption has changed over time as a response to availability and technological developments. Three distinct periods can be identified. The first period, extending from colonial times to about 1880, was characterized by the dominance of wood. The second period, from 1880 to 1920, was dominated by coal. And the third period, from 1920 to the present, is characterized by the use of a diversity of fuels—coal, petroleum, natural gas, nuclear energy, and renewable energy.

Dominance of Wood

Wood was the first major fuel consumed in the United States. Until 1850 wood supplied more than 85 percent of the total energy consumed in the nation. Waterpower was the second source of energy. Coal was of little importance. In 1850 bituminous coal supplied about 4.7 percent and anthracite 4.6 percent of the total. As late as 1875 wood supplied two-thirds of the energy consumed.

After 1875 the importance of wood as a fuel declined rapidly so that by 1900 it supplied only 20 percent of the total energy consumption and by 1925 only 6.8 percent. In volume of consumption, fuel wood reached a peak about 1870, when 138 million cords of wood were consumed. In Btu's the consumption of fuel wood rose from 2,138 trillion Btu's in 1850 to a peak of 2,893 trillion Btu's in 1870. During the same period the total consumption of mineral fuels rose from 216 trillion Btu's to 1,072 trillion Btu's.

The dominance of wood as a fuel for a period of more than 250 years was due to a number of factors. First, vast forests of the eastern United States made wood available. Although precise statistics are not available, it is known that firewood was consumed lavishly. As Reynolds and Pierson wrote in their study, Fire-wood Used in the United States, 1630–1930 (Forest Service Circular No. 641, Washington, D.C., 1942), "Cordwood was about as plentiful as air; but nobody wrote about it—why write about firewood, or even record statistics about it?" Second, even after coal was discovered, the deposits of the Appalachian Plateau were relatively inaccessible and far from major centers of population. Third, the energy from wood was quite adequate for most of the uses to which it was put. During the firewood era most of the fuel was consumed to heat homes. In the 1850s an estimated 90 percent of the firewood was burned in households. Per capita consumption was about 4.5 cords. An average family used about 17.5 cords of wood per year for home cooking and heating. This is the equivalent in heat of about 2.5 tons of coal per capita.

The relatively low demand for energy was a consequence of the period's limited industrial development. At the peak period of firewood production (about 1870) only 20 percent of this fuel was consumed by industry, about 800 trillion Btu's or the energy equivalent of 30 million tons of coal. Waterpower from streams provided most of the energy used in sawmills and gristmills.

During the nineteenth century the primary industrial use of firewood was in transportation. The first steamboat engines and the first locomotives were fired with wood. Railroads used wood as the principal fuel until the 1870s. Next to the transportation industries, the largest single industrial consumer was the iron industry. Around 1850, half of the iron produced was still smelted with charcoal. The quantity of charcoal consumed was relatively stable between the 1850s and 1870s, remaining at a level of 70 to 75 million bushels annually and rising to a peak of 86 million in the early 1880s.

After 1880 the consumption of wood declined rapidly due to two factors. First, the forest resources were approaching depletion. After 200 years of exploitation with no attempt to replenish the trees, the eastern United States forests were exhausted. Second, the development of anthracite and bituminous coal resources had grown so that coal was available to replace declining timber resources.

Trends in Energy Consumption

The consumption of energy in the United States has grown steadily upward. It has risen from 2.3 quadrillion Btu's in 1850 to 22.9 quadrillion Btu's in 1922 to 52.6 quadrillion Btu's in 1965 to a peak of 81.2 quadrillion Btu's in 1989. In recent years energy consumption has risen from 1.3 to 4.2 percent annually.

Since the mid-1960s energy consumption has surpassed domestic energy production. For example, in 1989 when domestic energy consumption totaled 81.2 quadrillion Btu's, domestic energy production was 65.87 quadrillion Btu's. The deficit is primarily in petroleum where domestic sources totaled 16.1 quadrillion Btu's while consumption totaled 34.26 quadrillion Btu's. In 1989 natural gas production totaled 17.2 quadrillion Btu's, but consumption totaled 18.6 quadrillion Btu's. In contrast, coal production exceeded consumption by 2.4 quadrillion Btu's.

Age of Coal

The first recorded commercial shipment of coal in the United States was reported in 1758 when 32 tons were sent from Norfolk, Virginia, by boat to the New York area. However, until 1830, coal production was not only small but also sporadic. In 1800 production totaled about 108,000 tons, and it was not until 1835 that production reached slightly more than 1,000,000 tons. In the next two decades growth was high, reaching 7,543,000 tons in 1855; a sound foundation was being established for the industry.

In 1855 more than three-fourths of the coal was mined in Pennsylvania, followed by Ohio with 600,000 tons, and West Virginia with about 350,000 tons. Coal emerged as an important fuel with the expansion of the iron and steel industries, the growing use of it in manufacturing, and the expansion of rail transportation. It is estimated that three-fourths of all coal produced in 1855 was transformed into steam power in mechanical works.

The shift to a mineral fuel in the smelting of iron ore began in eastern Pennsylvania where large supplies of anthracite were available. The transition from the hundreds of charcoal furnaces scattered throughout the nation was rapid. Anthracite furnaces were larger than charcoal furnaces and more capable of supplying the rapidly increasing demand for iron. In 1842 anthracite furnaces produced about 15 percent of the nation's pig iron, but by 1855 anthracite was used to produce about one-half of the pig iron, and in 1860 nearly two-thirds.

The use of bituminous coal for metallurgical purposes did not become important until the introduction of the Bessemer process in the 1860s. Bituminous coal is the only coal that can be converted into coke for the smelting of iron ore. The production of bituminous coal did not surpass anthracite production until 1870 when 20,471,000 tons of bituminous coal were produced as opposed to 19,958,000 tons of anthracite. In the period 1850 to 1880, when consumption of fuel wood increased about one-third, the production of coal increased about 950 percent.

After 1880 energy demand rose rapidly. Of the energy resources available—wood, hydropower, petroleum, and coal—only coal was available in sufficient quantity to fulfill the demand. In the period from 1880 to 1910 bituminous coal production doubled each decade, rising from 50 million tons in 1880 to 111 million in 1890 to 212 million in 1900 to 417 million tons in 1910. It reached its peak for the period in 1918 when 579 million tons were produced.

The expansion of anthracite production was at a slower but still significant rate, rising from 31.9 million tons in 1880 to a peak of 98.8 million in 1918.

Coal reached its relative peak of importance in 1899 when bituminous coal accounted for 71 percent of the energy consumed and anthracite an additional 18 percent, making a total of 89 percent. In 1899 petroleum and natural gas accounted for only 7.7 percent of the total. As petroleum production increased rapidly after 1900, the relative importance of coal declined. By

1918, the peak year of coal production, bituminous and anthracite supplied 82 percent of energy consumed, and petroleum and natural gas use had risen to 13.4 percent.

The expansion of bituminous coal production was primarily due to the demands of a rapidly growing industrial economy. By 1885 coal had gained acceptance as an industrial fuel. In 1885 when about 71 million tons were produced, about 42 percent was burned in locomotives, 13 percent made into coke, and the remaining 45 percent was distributed among all other industries and domestic users.

In the early twentieth century the demand for energy rose again. In the decade 1899 to 1909 the gross national product expanded by one-half and, in per capita terms, by nearly one-quarter. The output of manufacturing industries rose 58 percent. Railroad mileage continued to expand and electric power production grew rapidly. Basic pig iron and steel production rose 89 and 125 percent respectively. All of these industrial and technological developments created an increasing demand for energy which, in the prevailing opinion of the period, could be met only by increased utilization of coal.

Energy Diversification

The tremendous growth of energy demands in the twentieth century has required the utilization of all available resources. Consequently, the relative importance of each energy resource has changed as production varies. Most significant has been the decline of the relative value of coal. With increasing petroleum and natural gas production the relative value of coal as an energy source declined to 63 percent by 1929 and continued its decline to 26 percent in 1960. After 1960 U.S. coal production recovered somewhat, and by 1992 had risen to 32 percent of the energy consumption.

In contrast, petroleum's share of energy consumption has declined from a peak of 35.9 percent in 1960 to about 25 percent in 1992. Natural gas has followed the same general trend as petroleum, increasing from 3.3 percent in 1899 to 13.7 percent in 1945, 30.5 percent in 1960, but declining to 26.7 percent in 1992. Nuclear energy began in 1957 at Shippingport, Pennsylvania, and has risen to 6.9 percent of energy consumption. Waterpower production has increased many times, but has remained at about the same relative position: 2.8 percent of the total.

Energy Reserves

The energy resources of the United States are extensive but are unequally distributed and are highly variable in quality and accessibility.

COAL   Of the traditional fossil fuels, coal reserves are the largest. As of 1990, the Energy Information Administration reported that the demonstrated coal reserve consisted of 419.2 billion tons of bituminous, of which 312.1 billion tons were underground. Lignite reserve tonnage totaled 364.3 million tons, of which 319.3 were underground. In addition there were 7.3 billion tons of anthracite. Recoverability varies from less than 40 percent to more than 90 percent for individual deposits. About one-half of the coal reserves in the United States are estimated to be recoverable. There is thus several hundred years of coal reserves at the present rate of production.

PETROLEUM   The proved petroleum reserve has declined from a total of 33.6 billion barrels in 1977 to 27.9 billion in 1989. However, each year new reserves are added to the yearly total. Proved reserves in any field include both drilled and undrilled recoverable resources under economic and production systems now in operation. For a one-well pool where development has not gone beyond the discovery well, the area assigned as proved is small in regions of complex geological structure, but possibly larger where the geology is simple. In a sparsely drilled field the area between the wells is considered to be proved only if the information regarding the geology of the field and the production horizon is adequate to ensure that the unproven area will produce oil when drilled. Thus, the total increase of new oil, through annual discoveries, is comparatively small, but the total new oil proven by extension of known fields is comparatively large.

It is estimated that the ultimate recovery of petroleum in the United States will range from 33.2 to 69.9 billion barrels. The area with the greatest potential for finding undiscovered reserves is Alaska where the estimated range of new oil is from 3.6 to 31.3 billion barrels. The Gulf of Mexico has the second highest potential with an estimated range from 4.9 to 13.6 billion barrels. It is now estimated that the ultimate recovery of existing reserves of petroleum in the United States will be about 200 billion barrels of which 158.8 billion have already been produced.

NATURAL GAS AND NATURAL LIQUIDS The United States has an abundant reserve of natural gas. The proved recoverable reserve in 1989 was 175.4 trillion cubic feet. Production has totaled 777.8 trillion cubic feet with the ultimate recovery estimated at 953.8 trillion cubic feet. The greatest reserves are in Texas, the Gulf of Mexico, and Alaska. Natural gas liquids add an additional reserve of 7.8 billion barrels. This is the equivalent of 5.5 billion barrels of crude oil.

URANIUM OXIDE (U_3O_8) At the present time, atomic energy is developed primarily from uranium. The spectrograph and Geiger-Müller counter detect uranium in most granite and sedimentary rocks. However, uranium is rarely found in concentrated form, but ore containing a low content of uranium is practically limitless. The most important ores containing uranium include uraninite, pitchblende, davidite, carnotite, and autunite.

The largest uranium deposits in the United States are in the carnotite ores located on the Colorado Plateau of western Colorado, eastern Utah, northeastern Arizona, and northwestern New Mexico.

These ores were worked after 1910 to secure radium. In the late 1930s uranium was first recovered as a by-product of

Table 1.1
Known or Proved Recoverable Energy Resources

Fuel	Standard Units of Measurement*	Quadrillion Btu	Percent According to Btu Content**
Coal			
Anthracite	7.3	54.4	0.3
Bituminous	419.2	10,983.0	68.9
Lignite	45.0	603.0	3.8
Petroleum	27.9	161.8	1.0
Natural Gas	175.4	180.9	1.2
Natural Gas			
Liquids	7.8	36.0	0.2
Shale Oil	80.0	464.0	2.9
Uranium Oxide			
U_3O_8	1,537.0	3,458.2	21.7

*Coal in billions of tons; petroleum, natural gas liquids, shale oil in billions of barrels; natural gas in billions of cubic feet; and uranium oxide (U_3O_8) in million pounds.
**Reserves converted to Btu according to the following heat volume: anthracite, 12,700 Btu per pound; bituminous coal, 13,100 Btu per pound; lignite, 6,700 Btu per pound; petroleum and shale oil, 5,800,000 Btu per barrel; natural gas liquids, 4,620,000 Btu per barrel; natural gas, 1,032 Btu per cubic foot; and uranium oxide (U_3O_8), 450 billion Btu per ton.

vanadium. The ores contain 2 to 4 percent uranium oxide (U_3O_8). In the United States the reasonably assumed reserve of U_3O_8 is estimated to be 1,537 million pounds. If the much lower grade ores are included, the estimated additional reserves total 4,980 million pounds plus speculative reserves of 3,500 million pounds. These ores possess about 3,458 quadrillion Btu's.

SHALE OIL   There are vast shale oil deposits in the United States in an area of about 16,000 square miles in northwestern Colorado, northeastern Utah, and southwestern Wyoming. An initial survey by the U.S. Geological Survey in 1965 indicated a known reserve of 80 billion barrels. This included deposits yielding more than 25 gallons of oil per ton of shale in deposits 25 feet thick or more and limited to 1,000 feet below the surface. During the energy crisis of the 1970s some interest was shown in developing this reserve, but when petroleum became plentiful all projects were dropped. Potential reserves are still unknown, but initial estimates indicate that at least 18 billion barrels of oil exist in this region. This estimate is known to be far too low.

Petroleum

Oil is the world's most important source of energy and the most important single commodity in world trade. In addition, it is the most important raw material for the manufacture of chemicals. In the industrialized world oil is dominated by great corporations, and in the Third World the oil industry is controlled by national governments. The finances of the oil corporations and national oil industries dwarf the national budget of all but the largest countries. The profits of the oil corporations may exceed \$2 to \$3 billion and the estimated cost of the industry's property, refineries, and equipment is over \$300 billion.

Production and Consumption

The modern oil industry began with the discovery of oil at Titusville, Pennsylvania, in 1859. For more than a century the production of oil in the United States increased. In 1909, the fiftieth anniversary of the industry, U.S. production reached 500,000 barrels a day—more than the rest of the world combined. The demands for oil continued to increase as more diverse markets developed.

One of the first indications that the United States oil industry could not continue its growth indefinitely came in 1948. In that year the nation imported more oil than it exported. For the first time the United States had become a net importer of oil. Nevertheless, the nation continued to produce half of the world's oil in the early 1950s. In addition the United States had sufficient unused capacity to produce for increased world demand in an emergency, as it did in 1956 during the Suez crisis.

In the 1950s U.S. oil became more expensive to produce than in many other areas of the world, particularly the Middle East. As a response, the U.S. government began to place restrictions on imports in order to maintain national security and to protect the powerful domestic oil companies. Due to the resulting high oil prices, domestic oil production continued to rise.

In the 1960s as demand continued to grow, cheap imported oil continued to rise, commanding an ever-larger share of the U.S. market. The historic turning point came in 1970 when U.S. spare capacity vanished and U.S. production reached its ultimate peak— an average of 11.3 million barrels a day. Since then the level of oil production has steadily declined. Oil imports account for 40 to 50 percent of annual consumption in the United States. In 1990 the average daily consumption was 12 million barrels.

Petroleum Provinces

Petroleum production is very widely distributed in the United States. Although production occurs in ten major provinces in 34 states, most of the production occurs in a few states.

Appalachian Province

The modern world oil industry began in the Appalachian Province where the first well was drilled specifically for oil at Titusville, Pennsylvania, on August 27, 1859. This region extends from southwestern New York through western Pennsylvania and eastern Ohio to Tennessee. The Appalachian fields were the only suppliers of the nation's oil until the discovery of the Lima-Indiana, fields in 1884. The maximum production in the Appalachian fields was reached in 1891, when more than 33 million barrels were produced. By the 1890s it was recognized that the fields were being depleted and oil men, who a few years before produced oil without regard to conservational practices, now realized that the future looked bleak. One of the greatest contributions to the oil industry was the realization that an oil field could be rejuvenated by means of water flooding the oil sands. Because most of the oil fields were extremely small, they did not lend themselves to secondary recovery of oil by water flooding. Only the Bradford field in McKean County, northern Pennsylvania, had a sufficiently large area of uniform oil-producing sand that was suitable for this procedure. Oil production increased until 1937 when a second peak of 37 million barrels of oil was reached. Since then production has once again declined to about 10 million barrels annually.

Although production is small, a number of factors favor the continuance of the tiny Appalachian oil industry. Because these fields produce the finest lubricating oil, the crude oil commands a premium price. Trade names of Pennsylvania companies such as Quaker State, Pennzoil, Kendall, and Valvoline are known throughout the world. Small refineries exist in the region to process local production. However, because production is small, refineries must import oil from other oil areas.

There are also other considerations. Once a well is drilled and placed in production, maintenance costs are minimal and production can continue even if it is only a tiny output per day. The landowners also encourage continued production, for most receive royalties of one-eighth of production. In a depressed economic region, the small oil income is a welcome bonus. Finally, proved recoverable resources in the Appalachian Province total more than 80 million barrels. While this region is of no importance in providing the energy requirements of the nation, small production will continue for many decades.

Lima-Indiana Province

The region extends from the northwestern corner of Ohio to Indianapolis, Indiana. Oil was discovered in 1884, and peak production was attained in 1896 in Ohio, and in 1906 in Indiana. Because of the failure to apply conservation principles in the region, production rose and declined with remarkable rapidity. Although more than 60,000 wells were drilled in this area, onetenth of this number would have recovered more petroleum. The province has long been exhausted.

Michigan Province

The Michigan Province was discovered in 1925, and peak production occurred in 1939. After a period of decline, secondary recovery methods were applied and production increased. In recent years the province has had an annual production of about 30 million barrels.

Illinois Province

This oil region lies in southeastern Illinois and western Indiana. Oil was first discovered in 1905 and an initial peak was reached in 1910 when 33 million barrels were produced. Production had declined to less than 5 million barrels by 1937, when deep drilling discovered new sources of petroleum. Output increased rapidly to a second peak of 147 million barrels in 1940, but has subsequently declined, and has varied from 22 to 30 million barrels in recent years.

The two peaks and the rapid decline in the production curve reflect the lack of conservation legislation to control output. Because conservation principles were ignored, the ultimate yield from the Illinois field has been greatly decreased.

Gulf Coast Province

The famous Spindletop well, drilled in 1900, was the first major gusher in the United States. Within a week hundreds of thousands of barrels of oil flowed from this single well. This well initiated the massive production of oil in the United States. The Gulf Coast Province, with fields on coastal Louisiana and Texas, for decades annually produced between 500 and 600 million barrels of petroleum. However, in recent years production has declined to between 300 and 350 million barrels. The oil is usually associated with domal structures in which the oil is concentrated at the top of the dome. Thus, this oil is frequently under great pressure, producing many gushers. The crude oil is of a heavy asphaltic base. Fuel oils are the major refined product. However, with modern refinery processes, the heavy oils are converted into lighter products, such as gasoline.

Midcontinent Province

The Midcontinent Province includes the oil fields of Kansas, Missouri, Arkansas, Oklahoma, northern Louisiana, northern and western Texas, and eastern New Mexico. This vast petroleum area is the most productive petroleum region in the United States. Production began in 1905 and rose rapidly. Since 1920 annual production has ranged from 500 to 1,200 million barrels. Recent production totals about 800 million barrels per year.

All grades of petroleum are produced in the region, from light paraffin based to heavy asphalt based oils. This province has many oil fields, including the prolific east Texas field. Although there was great waste in the past, conservation is now being practiced. Secondary and tertiary recovery are of major importance in maintaining the production of this region.

Northern Great Plains Province

This was the last petroleum province to be discovered in the conterminous United States. The major area of production is the Williston Basin of North Dakota, eastern Montana, and the northern part of the provinces of Saskatchewan and Manitoba in Canada. Exploration and development have been active since the discovery of oil in 1951. Proved reserves total about 1 billion barrels. Although lack of a local market and long distances to consuming centers have retarded development of the region, its annual output in recent years has been about 60 million barrels.

Rocky Mountain Province

This province is composed of scattered fields from Montana to New Mexico. Production began in 1916 but, due to distance to markets, output increased slowly. Intensive exploration began about 1940, and many new discoveries have been made. Wyoming is now the seventh major petroleum-producing state, with an annual output of about 115 million barrels. Most of the oil is marketed by a pipeline system that serves the eastern United States.

California Province

The oil fields of California are the third largest producers in the United States. Production began in the area in 1887, and California has stood between the first and third producing states since then, exceeded only by Texas and Alaska. In recent years annual output has been about 350 million barrels. Most of the oil production comes from the Los Angeles Basin and the southern portion of the San Joaquin Valley. The oil pools are characterized by high gas pressure so that individual wells are normally large, since the gas is the motivating force moving the oil to the well. Because of flush production and competition, excessive drilling has been common in this area. Conservation laws now control production in a program designed to prevent future occurrences of the gross wastefulness of the past.

Alaska Province

The last major oil field was discovered in northern Alaska at Prudhoe Bay in the late 1960s. It proved to be one of the nation's major fields with annual production of over 300 million barrels. To market the oil, a pipeline was laid across Alaska to Valdez on the southern coast where tankers transport the oil southward to the West Coast of the United States. Vast areas in northern Alaska are potential oil producers. This land is now held as a reserve for future exploration.

American Dominance of the World Oil Industry

At about the turn of the twentieth century, U.S. oil companies began to seek oil concessions in Mexico and other countries. At the end of World War I the United Stated produced two-thirds of the world's oil and another sixth from U.S. companies operating in Mexico. Because the demands of World War I had greatly reduced U.S. known reserves, the government became concerned about maintaining adequate oil supplies. Major new discoveries were not occurring in the United States and for a few years the country became a net oil importer.

The acquisition of foreign oil concessions was actively sought. The policy developed that the U.S. government would not purchase the oil concession directly because the U.S. Department of State felt that foreign oil controlled by U.S. companies was as reliable as if the foreign oil was owned by the U.S. government. It then became policy that the U.S. government and the U.S. oil companies cooperated to acquire as many foreign sources of oil as possible. In 1920, Congress passed a law that prohibited foreign-owned corporations from acquiring oil leases in the United States public lands if their government did not allow U.S. firms to explore for oil in territories they controlled. The law was aimed primarily at Royal Dutch Shell who persuaded the Dutch to open the Netherlands East Indies to exploration by U.S. companies.

In the period from the 1930s to 1950s the United States gained a dominant position in the exploitation of Middle East oil fields. Of major importance, the U.S. government obtained an open door for investments in the traditional British and French zones of influence, including Iraq. As a result Exxon and Mobil acquired about one-quarter of the Iraq Petroleum Company. In response to these developments Iraq in 1934 became the second major Middle East exporter, exceeded only by Iran.

The U.S. government in 1934 persuaded the British government to allow Gulf Oil to enter Kuwait, a British protectorate. British Petroleum and Gulf Oil entered a cooperative arrangement with the sheik of Kuwait to develop the oil reserves and avoid competing against each other. The U.S. government also obtained British approval for Standard Oil of California (SOCAL) to enter Bahrain in 1932, a small British protectorate off the coast of Saudi Arabia. In 1933 SOCAL secured the right to begin oil exploration in Saudi Arabia. A few years later SOCAL sold a half interest in its Saudi Arabian rights to Texaco, thus forming the Arabian American Oil Company: Aramco.

In order to protect U.S. interests in Saudi Arabia the U.S. government responded to the needs of the nation. For example, during World War II when the annual pilgrimages to Mecca were reduced, Aramco told President Roosevelt that King Ibn Saud needed $6 million quickly. Washington responded by providing lend-lease assistance. In 1945 the U.S. National Security Council stated, "It is in our national interest to see that this vital resource [Saudi Arabian oil] remains in American hands."

In 1946 when SOCAL and Texaco decided to sell part of Aramco to Exxon and Mobil, antitrust questions were raised. In order to expedite the sale, the U.S. government provided antitrust clearance, indicating the larger consortium was in the national interest. King Ibn Saud agreed to the transactions, which meant faster development and greater money for his nation. The United States also pressured the British and French governments to overlook clauses in the Iraq Petroleum Company agreement that would have prevented Exxon and Mobil from buying in.

In 1950 Saudi Arabia demanded additional money from the companies. This created a number of problems. The companies did not want to raise oil prices for their European customers because this would have reduced the postwar recovery and the effects of the Marshall Plan. Nor did the companies want to raise royalty payments to Saudi Arabia since this would have reduced their profits. The issue was resolved when Aramco began to make payments to the Saudi Arabian government in the form of income taxes owed the U.S. Treasury—a tax credit. Tax credits became a common practice for other U.S. companies operating abroad. Nevertheless, the U.S. Department of State expressed some concern over what, in effect, amounted to a company subsidy to Saudi Arabia paid by U.S. taxpayers.

Iran was another area where Aramco interests were significant. In 1946 the United States put pressure on the Russians to evacuate northern Iran and keep them from threatening the Middle East oil interests. In 1951 the Iranians, led by Prime Minister Mohammed Mosaddegh, nationalized British Petroleum's Iranian properties. The seven major oil companies—Exxon, SOCAL, Texaco, Mobil, Gulf, British Petroleum, and Royal Dutch Shell (60 percent owned by the Dutch)—immediately instituted a boycott of Iranian oil. These seven companies controlled 98 percent of the world's oil market and easily increased production to replace the lost Iranian oil in the world market. The U.S. government supported the oil companies for fear that if Mosaddegh succeeded, other major oil-producing countries might follow suit.

In response to U.S. activities, the Mosaddegh government fell and the Shah was reinstated. The U.S. government assumed the lead in settling the dispute, in which the U.S. companies gained a new consortium of companies to operate Iran's oil fields and market its crude. While some of the oil companies were reluctant to join the consortium, the U.S. government wanted U.S. oil companies to participate because they were the only companies capable of producing sufficient oil to provide the income deemed necessary for the new Iranian government to function. The first ownership in the consortium was apportioned as follows: U.S. companies, 40 percent; British (British Petroleum and Royal/Dutch Shell), 54 percent; and French (Compagnie Française des Petroles), 6 percent. The Iranians resented the settlement because foreign companies continued to dominate their oil industry.

The formation of the Iranian consortium represented the zenith of the control of foreign oil resources by U.S. oil companies. U.S. influence in oil production was felt in other areas of the world. In Venezuela, U.S. and British interests developed the oil fields as the Dutch had done in the Netherlands East Indies.

Organization of Petroleum Exporting Countries

Oil Company Dominance

By the late 1950s the oil exporting nations believed that oil profits were going to the oil companies. While oil prices had risen on the world oil markets the "posted price," that is the official price used to determine host-government revenue, remained constant. As was the custom of the period, the oil companies did not negotiate with the oil-exporting nations. The price of oil was determined unilaterally. The helpless oil-exporting nations were forced to accept the oil price decline.

In response to the cuts, Iraq invited Iran, Kuwait, Saudi Arabia, and Venezuela to form a permanent organization called the Organization of Petroleum Exporting Countries (OPEC). The fundamental purpose of the organization was to create a permanent group of exporting countries to coordinate and unify the policies of its members. The first resolution expressed the views of the member countries that they could no longer remain indifferent to the attitude heretofore adopted by the oil companies in affecting prices.

The creation of OPEC was widely recognized as the beginning of a cartel. In the 1960s OPEC's influence on the policies of the oil companies was minimized. It was a period of excess oil and each of the exporting countries wanted to maintain production in order to receive needed revenues. The member countries of OPEC met regularly to assess the oil market and price structure and provide basic information to its members. In the first decade, OPEC doubled its membership by accepting Qatar in 1961, Libya and Indonesia 1962, Abu Dhabi in 1967 (in 1974 Abu Dhabi became a part of the United Arab Emirates), and Algeria in 1969. During the second decade three new members were added: Nigeria, 1971; Ecuador, 1973; and Gabon in 1975.

OPEC Control

The late 1960s and early 1970s were, for the most part, years of high economic growth for the industrial world. The growth was fueled by oil. As a consequence, the 1970s witnessed a dramatic growth in world oil demand. The 25-year oil surplus ended with demand exceeding available supply. Free world oil demand rose from 19 million barrels per day in 1960 to more than 44 million barrels per day in 1972.

In the late 1960s the U.S. domestic oil industry could no longer meet demand. For decades U.S. oil production had been regulated by the Texas Railroad Commission, the Oklahoma Corporations Commission, the Louisiana Conservation Commission, and similar bodies in other states. These bodies controlled output, keeping actual production well below capacity in order to promote conservation and maintain prices in a situation of chronic potential oversupply. The results of these endeavors were to provide the

United States and the entire Western World with a security reserve that could be called upon in times of crises, such as the major extended demand of World War II and the much more limited crises of 1951, 1956, and 1967.

The oil industry in the United States changed drastically in the 1960s. In the period 1957 to 1963, U.S. surplus capacity had totaled about 4 million barrels per day. By 1970, less than 1 million barrels per day remained. That was the year of peak production with 11.3 million barrels per day. From then on, oil production began to decline. In March 1971, for the first time, the Texas Railroad Commission allowed production at 100 percent capacity. With consumption continuing to rise, the United States had to turn increasingly to the world oil market to satisfy domestic demand. The import quotas, originally established by President Eisenhower, were removed, and net imports rose rapidly from 2.2 million barrels per day in 1967 to 6 million barrels per day in 1973. Imports as a share of total oil consumption rose between 1967 and 1973 from 19 percent to 36 percent of the total consumed.

The disappearance of surplus capacity in the United States had major implications, for it meant that the "security margin" upon which the Western World had depended was gone. In November 1968, the State Department informed the European governments at an OPEC meeting in Paris that U.S. production would soon reach the limits of its capacity. The European nations were taken by surprise that the United States' oil supply was declining.

As a response to the decline of U.S. production, there was an increasing reliance on Middle East oil and such new areas as Nigeria. Between 1960 and 1970 free world oil demand had grown by 21 million barrels per day. During the same period, production in the Middle East (including North Africa) had grown by 13 million barrels per day. Thus, two-thirds of the large increase in oil consumption was being secured from wells in the Middle East.

A basic goal of OPEC was to increase the revenues from the oil companies. Libya, under the new regime of Colonel Moamar Qaddafi, was the first OPEC nation to challenge the oil companies. The process started with the nationalization of the Chappaque Oil Company, a U.S. company alleged to have bribed a number of officials. Libya followed this united move by ordering reductions in production of all foreign oil companies. For example, the output of the Occidental Oil Company was reduced from 845,447 to 464,565 barrels a day. This move created an atmosphere of fear in the oil industry. Rumors of imminent nationalization of all oil resources were widely circulated.

The Occidental Oil Company was the first to come to an agreement with the Libyan government. The agreement increased the posted price by 30 cents from $2.23 to $2.53 for crude. The price was to increase by 2 cents per year for a period of five years. To encourage other companies to agree, the Libyan government increased the production of Occidental from 464,565 to 700,000 barrels daily. Within two months all other companies developed agreements with Libya.

As a response to the Libyan actions, the OPEC minister at the twenty-first conference in Caracas, Venezuela, in 1970 spelled out the following objectives:

1. Establish 55 percent as the minimum rate of taxation on the net income of the oil companies operating in member countries
2. Eliminate existing disparities in posted or tax reference prices of crude oil in member countries on the basis of higher posted prices applicable in the member countries
3. Establish a uniform general increase in the posted price or tax reference price in all member countries to reflect the general improvement in the conditions of the international petroleum market
4. Adopt a new system for the adjustment of gravity differences of posted or tax reference prices
5. Eliminate completely the marketing allowance granted the oil companies from January 1, 1971

This resolution showed the determination of OPEC countries to gain control of their domestic oil production. Over the next two years the OPEC nations increased prices modestly. It was, however, the Arab-Israeli War of 1973 that triggered the rise in oil prices and the ultimate control by OPEC of its oil resources. With tight market supplies, crude oil prices began to increase. On October 15, 1973, the oil ministers met in Kuwait and unilaterally increased the posted price from $3.011 to $5.119 per barrel.

Oil Embargo

After the oil ministers increased prices, OPEC decided to use the "oil weapon" in support of the Arab cause, mainly because the

United States supported Israel in spite of the strong appeal from Arab leaders, especially King Faisal of Saudi Arabia. The oil minister reduced oil production by 5 percent, about 900,000 barrels a day. Thus consuming countries were divided into three categories:

1. Friendly nations   Those who opposed the Israeli occupation of Arab territories; mainly France, Britain, Spain, and eastern European countries
2. Non-friendly nations   Those who supported Israel, namely the United States and Netherlands
3. The rest of the consuming countries were allowed to secure oil from the Arab countries

As a response to production cutbacks and restrictions on exports, Secetary of State Henry Kissinger remarked that the new situation "altered irrevocably the world as it had grown up in the postwar period."

Dominance of OPEC

The Arab oil embargo did not last long. The Arabs could not maintain their position while faced with mounting world criticism. In the long run, far more important to the United States, OPEC became the unchallenged power in determining the price of crude oil. As a response, crude prices rose from $5.12 in October 1973 to $11.65 by January 1974. From 1974 to 1978 oil prices remained nearly stable, rising to $13.30 in July 1977. During the 1970s and early 1980s the OPEC nations nationalized their oil. The output of petroleum was then determined by OPEC and not the United States and other foreign companies.

In 1975 the revolution in Iran reduced oil production, creating uncertainty in the international petroleum market. This situation led to an increase in the demand for OPEC oil and to an increase in price. The established price of oil went from $12.70 to $14.54 a barrel. Spot prices rose more rapidly and by 1979 reached $24.00 to $26.00 a barrel. Each country decided on a general sale price which it thought could be supported in the market. For example, Libya raised its official price to $34.67 a barrel.

As a response to the second dramatic rise in oil prices, member countries of OPEC began to realize that their lack of coordination weakened the organization and led to pricing confusion and inconsistencies. They also felt that the return to a crude oil pricing formula was important in stabilizing the international oil market. Consequently, at the 57th OPEC Conference in June 1980, OPEC agreed to set the price level for marker crude (crude oil with a given quality that sets the industry standard) at a ceiling of $32.00 per barrel. Price confusion continued and the spot price reached $41.00 per barrel.

In the years 1973 to 1980 OPEC became the unchallenged power in determining the price, the royalty rate, the income tax and, most important, the percentage of the foreign oil companies each country wanted to acquire, including complete nationalization of the industry. The oil companies had lost complete control of oil production. The price of oil became chaotic.

Decline of OPEC

In the 1980s the world's oil consumption decreased. In 1979 world consumption of oil was 64.1 million barrels a day, but by 1980 the demand was but 61.7 million barrels daily. The world decline was not only due to an economic recession but also the development of alternative energy sources, particularly coal and the increase of oil production in non-OPEC nations.

In the early 1980s the OPEC nations attempted to adjust to the new market conditions. Quotas were established, but rarely adhered to by the OPEC nations in their attempts to maintain their oil revenues. Oil prices declined to a level of $17 to $18 a barrel.

The decline of OPEC's influence resulted from their failure to grasp the true dimension of the international petroleum market and, at the same time, coordinate their own efforts. Internal disagreements weakened OPEC. The countries were not able to recognize that when one member suffered other members would suffer sooner or later. Politics also contributed to the decline of OPEC. The oil ministers had very little power to decide important oil issues, and the heads of state looked on oil as one of the cards they played for or against members depending upon the nature of the countries' relationship at any given time.

Response of the United States to OPEC

Initial Response

In the early 1970s there was some recognition that the petroleum industry was experiencing a number of far-reaching changes. The peaking of domestic products and the gradual increase in international prices made the domestic control of production seem unnecessary. Accordingly, state proration agencies gradually raised "allowables" to 100 percent and, in 1971, the federal government abolished the mandatory import control system and instead set a small fee for imports. The Nixon administration's price regulations created some periodic shortages of petroleum products, notably heating oil. Shortages and distortions in the distribution of crude oil and its products in turn led to growing demands for governmental allocation of supplies. Thus the general economic policies of the early 1970s added to the complex controls that affected the energy system of the country.

Energy problems did not begin with the oil embargo of 1973, but only then did the general public become aware that oil could become a scarce commodity. There was now widespread recognition that the stakes were high and a crisis existed. The problem was not solved by the implementation of a single plan but rather by a series of steps each of which, for lack of broadly acceptable alternatives, eventually became a key element in the national solution.

Nixon Response

Just three weeks after the oil embargo, President Nixon delivered his first energy message to Congress in which he outlined a broad plan, named Project Independence, to make the country energy self-sufficient by 1980. Because the plan had little careful analyses, it set goals that had no chance of realization. The speech did set the tone for what was to become the basic Executive Branch approach to the oil crisis. The threat to the economy and the nation's security required the government to intervene. As a response, price controls, allocation of supplies, entitlements, and the enlargement of the government's regulatory apparatus—all supposedly temporary expedients to manage shortages—quickly became institutionalized.

The embargo and price increases were not only of concern domestically, but created some important foreign policy problems for the United States. The first reaction of most European governments was to protect their own interests. This created a round of bitter charges and counter charges. In order to control the international situation, Secretary of State Kissinger attempted to reassert U.S. leadership. His solution was to create an association of industrialized oil-importing countries. In February 1974, the International Energy Program (IEP) and the International Energy Agency (IEA) were formed to blunt the effectiveness of future embargoes. Twenty countries joined agreeing to: (1) create stockpiles of oil equal to 60 (later 90) days of net imports, (2) have "demand restraints" policies for use in emergencies, and (3) follow a formula for sharing oil among themselves in case of an embargo or other emergency.

The success of the international control of energy was not matched at home. During the Nixon administration the passage of the Emergency Petroleum Allocation Act (EPAA) was most important. In the name of equity, it set in place a host of controls over petroleum prices, production, and marketing. The basic goal of the EPAA was fairness, that is to distribute short supplies equitably among those consumers that depended on them, and to protect consumers and the economy as a whole. This program was essentially designed to solve immediate problems with little consideration for long-range trends.

Ford Response

When Gerald Ford assumed the presidency he sought to reassert executive leadership in the energy field. In his January 1975 State of the Union address he assured the nation he was going to reformulate energy policies in order to relate market forces with energy independence. He presented Congress with a long list of new legislation including deregulation of natural gas, excise taxes on imported crude oil and products, creation of a strategic petroleum reserve of a billion barrels, and creation of a $100 billion Energy Independence Authority. Ford's plan never made it out of legislative committees.

Under Nixon and Ford, solving the energy problem was a response to a series of incremental actions, some taken by regulators, some by Congress, but all influenced by the existing patterns of energy regulations and political pressures due to imposed international changes.

Carter Response

When Jimmy Carter became president he made energy a first priority. The energy situation appeared to have worsened in that oil imports were up about 31 percent since 1973. In response, a document that came to be known as the National Energy Plan was developed. While it included much from previous plans, it differed in three important respects. First, it stressed the goal of ensuring equity in the distribution of benefits and burdens. Second, it emphasized the importance of conservation, treating this concept as a source of new supply. Finally, the plan recognized the long-term nature of the energy crisis. Recognizing the growing dependence on foreign oil, the plan stressed the importance of shifting the country to the use of coal and other alternative sources of energy.

Congress took a year and a half to pass a greatly changed Carter Plan. The crude oil equalization tax was gone and the gas compromise, originally based on regulated prices tied to fuel oil, had become a program of postponed deregulations. In reality the heart of the program, the mechanism by which costs and benefits in energy would be allocated, was omitted. Congress did pass legislation dealing with coal conversion, utility reform, and the promotion of conservation. Among the most important was the gas-guzzler tax, and a subsidy for increased production of gasohol. Basic renewable energy legislation was also passed.

With the second oil shock of 1978–1979 the nation's energy situation again deteriorated. Gasoline lines and a crisis atmosphere reappeared. On April 5, 1979, Carter again introduced a new package of energy-solving proposals. In order to increase domestic oil production he proposed to decontrol the price of domestic oil but also called for a Windfall Profit Tax in order to protect low-income groups. In the ensuing months a number of stopgap measures were passed to survive the immediate energy crisis. The last energy bill to be enacted under the Carter administration was the Energy Security Act known as the Synfuels Bill. It created a Synthetic Fuels Corporation to develop synthetic fuels from coal, shale, or tar sand. After billions of dollars of expenditure the synfuel program was discontinued in the late 1980s.

Modern Era

With the increase in the world production of oil in the early 1980s the Reagan administration promoted conventional energy production, attempted to remove regulations and legal obstacles, and enlarged access to energy resources. One major endeavor was a new oil land leasing system designed to make nearly one billion acres of public lands—mostly offshore—available for private lease to oil and gas companies. In reality only a tiny fraction of the public lands were leased.

When the Emergency Petroleum Allocation Act expired in October 1981, Congress renewed the act. However, Reagan vetoed it, indicating that there was no need to impose price controls and allocate supplies of crude oil products by utilizing standby rationing plans.

In the period after 1980 there have been no disruptions of oil severe enough to tighten supply or send prices soaring. Despite the extended war between Iraq and Iran and the continuing tension between Israel and its Arab neighbors, oil has continued to flow. However, the importance of Middle East oil availability remains a critical issue. When Iraq invaded Kuwait in 1990, Operation Desert Storm ensued with the United States driving Iraq from Kuwait. The United States recognized that the huge oil reserves of Kuwait could not be held by an unfriendly nation.

Future

The United States remains the world's largest petroleum consumer. About half of the petroleum consumed must be imported. No new major discoveries have been made in the United States since northern Alaska more than 20 years ago. American society remains one where the automobile plays a dominant role in providing personal transportation. There is no doubt that the massive trade deficits of the nation are primarily due to the voracious need for petroleum. No other alternative source of energy to replace petroleum is on the horizon. For the foreseeable future petroleum will continue as a major source of energy for the nation.

Bituminous Coal

Coal has played a significant role in the development and industrial progress of the United States. Historically, coal was a primary fuel used to build the United States into a great industrial nation. While coal is no longer the single greatest energy source, it remains indispensable producing nearly 60 percent of the nation's electric energy, and about one-quarter of the total energy consumed. Since the early 1970s when U.S. oil production declined and imports rose, coal has played a greater role in the nation's energy economy, increasing production by about 80 percent since 1971. At the present time coal contributes about $21 billion annually to the economy. As the United States moves toward a more electrified economy, coal is expected to play a vital role in meeting this increased demand for energy. The development of innovative, clean, coal burning technologies will help meet this demand in an efficient and cost-effective manner, while protecting the environment.

Coal Provinces

The nation's coal resources are extensive but unevenly distributed and highly variable in quality and accessibility. The coal areas have been classified into six major provinces: Appalachian, Interior, Gulf, Northern Great Plains, Rocky Mountains, and Pacific Coast.

Appalachian Province

The Appalachian Province is the oldest coal mining region in the nation. It is divided into two parts: The anthracite fields of northeastern Pennsylvania produce about 99 percent of the nation's anthracite, and the bituminous fields extend from north-central Pennsylvania to central Alabama. The bituminous coal beds were formed in great swamps that range in geologic age from Pennsylvanian to Permian (260–300 million years ago). Deposition of vegetation in the coal-bearing Pennsylvania Epoch persisted uninterrupted for a period of about 50 to 60 million years. The geologic structure throughout the coal fields is characterized by nearly horizontal to gently undulating broad northeast-southwest trending formations.

Over 50,000 square miles are underlaid by bituminous coal with an estimated reserve of 107 billion tons, of which 86 billion are underground. West Virginia has the largest reserve with about 30 billion tons followed by Pennsylvania with 29 billion tons. In 1990 the Appalachian Province provided 443 million tons or about 48 percent of the nation's coal output. The coal-bearing rocks have a thickness of 1,200 to 3,000 feet. Major rock types are shale, sandstone, limestone, claystone, and, of course, coal. There are more than 50 named coal seams in the Appalachian area. While a majority of the coal beds have been mined locally where seam thickness and quality enable profitable operations, most of the coal produced is from about a dozen major seams. Of the coal seams, the Pittsburgh is the most important. Lying in southwestern Pennsylvania and adjacent areas it varies from 6 to 10 feet in thickness and is an excellent cooking coal. It has been intensively mined since the middle of the nineteenth century, and is now nearing exhaustion.

Coal quality in the Appalachian field varies both locally and regionally. The coal beds exhibit a progressive change in fixed carbon, volatile matter and calorific value from east to west. Regional metamorphosis was greatest in the eastern margins of the field. Consequently the rank of coal changes from east to west; low volatile bituminous (fixed carbon content between 78 and 86 percent) to medium volatile bituminous coal (fixed carbon content between 68 and 78 percent) and high volatile bituminous coal (fixed carbon content less than 68 percent). Regrettably, many of the Appalachian coals are high sulfur (1 to 4 percent) and have a moderate high ash content (8 to 15 percent). With the enactment of the Clean Air Act of 1970 many Appalachian coals cannot be burned until the sulfur content has been reduced. The heat value of the coal normally ranges between 10,000 Btu/lb to 14,000 Btu/lb.

Interior Province

The Interior Province is divided into four regions: eastern, western, northern, and southern. The coal of the Interior Province is largely bituminous, but not of as high a quality as that of the Appalachian Province. Of the four regions, the eastern, comprising Illinois, Indiana, and western Kentucky, has been mined most intensively. In 1990 production totaled 106 million tons of which about 75 percent came from Illinois. Illinois, with 77 percent of its area underlaid with coal, has the greatest reserves of any state east of the Mississippi River. The coal resources of the western region, extending from Iowa to Arkansas, are large, but consist mostly of low-rank bituminous coal. Mining is limited, with an output of 4 to 5 million tons annually. Sulfur content of the coal is high, which led to decreased production in recent years. Only limited mining occurs in the northern and southern regions.

Gulf Coast Province

The lignite region of the Gulf Coast extends from Texas to Louisiana. The coal is in a region of young, undistributed rocks so that it has remained low rank. Because of its low heat value, exploitation was long delayed. Texas has reserves of 13,500 million tons and Louisiana an additional 500 million tons. Only since 1975 has significant production developed. This province now produces over 50 million tons of lignite annually.

Northern Great Plains Province

The Northern Great Plains Province includes the coals of the United States' Great Plains and extends into the Prairie Province of Canada. This province has a tremendous reserve of about 166 billion tons of subbituminous coal and 25 billion tons of lignite. Montana has the largest reserve of subbituminous with about 62 percent of the total, followed by Wyoming with 37 percent of the total. Of the lignite deposits, 60 percent are in Montana and 30 percent in North Dakota. About half the reserves of coal can be mined by surface operations. There was little exploitation of these coal reserves until after the passage of the Clean Air Act of 1970. This coal is essentially free of sulfur. With the development of the unit trains that move as many as 200 cars, the cost of transportation is no longer prohibitive. These trains are loaded at the surface mine and move to their destination in the Midwest without stopping. Production has risen from about 10 million tons in 1970 to over 200 million tons of subbituminous coal in recent years, of which about 75 percent comes from Wyoming.

Rocky Mountain Province

The Rocky Mountain Province is composed of a large number of isolated fields extending from Montana to New Mexico with a reserve of 28 billion tons of coal, of which about 60 percent is bituminous, 35 percent subbituminous, and 5 percent lignite. Colorado has the largest reserve with about 70 percent of the total followed by Utah, New Mexico, and Arizona. Coal production has been about 60 million tons annually with New Mexico having 30 percent of the output, Utah at 25 percent, and Colorado at 23 percent.

Pacific Coast Province

The coal of the Pacific Coast is limited to the state of Washington, although there are mine deposits in Oregon and California. The coal reserves of Washington total about 1.4 billion tons, of which about 70 percent is subbituminous. Mining is difficult because of folding and faulting, and annual output totals between 4 and 5 million tons.

Ranks of Coal

Coal is a combustible earth material composed of fixed carbon, hydrocarbons (volatile matter), moisture, impurities (sulfur), and ash originating from the alteration of plant life. Coal develops from vegetation that has formed in swamps, making peat bogs. Peat develops into coal in a number of stages. It is estimated that it takes 100 years to form 1 foot of peat and from 3 to 8 feet of peat to produce 1 foot of coal. Pressure, therefore, is a major factor in coal formation. Lignite is characteristically found where burial of the peat is shallow and, consequently, little pressure is applied. In contrast, anthracite and other high-carbon coals are formed by great compression related largely to crustal disturbances.

The rank of coal refers to the differences in the progressive evolution of coal from lignite to anthracite. The changes are characterized by a decrease of volatile matter and an increase in the fixed carbon. The higher the fixed carbon, the higher the ranking. This classification, which is based on chemical and physical properties, was devised by the United States Geological Survey.

Lignite, commonly known as brown coal, is the lowest in rank. It has a fixed carbon content of 30 to 55 percent and volatile matter from 18 to 20 percent. The structure is fibrous and woody. It has a moisture content of 30 to 43 percent, and disintegrates readily on exposure.

Subbituminous coal is black and may be dull or shiny. Fixed carbon varies from 35 to 60 percent and volatile matter from 30 to 40 percent. The moisture content ranges from 2 to 40 percent. On weathering it crumbles readily and has a tendency toward spontaneous combustion. The heat content of this coal is fairly high.

Bituminous coal is the most important industrial and heating coal in the world. It may be dull black to highly lustrous. The fixed carbon content varies from 48 to 73 percent and the volatile matter from 32 to 42 percent. The moisture is low, varying from 3 to 11 percent. This coal stores well and burns with a yellow flame. There are many varieties, of which the most common are coking, noncoking, cannel and boghead.

Semibituminous coal has a fixed carbon content of 73 to 80 percent and volatile matter of 20 to 25 percent. The coal burns with little smoke and has the highest heating value of all ranks of coal. The coal has less than 3 percent moisture content. It is a rich black, usually with a high luster. Its greatest handicap is its friability, which causes it to crumble easily. Semibituminous is an excellent steam coal and is used for general heating purposes.

Semianthracite differs from anthracite primarily in that it is more friable. It has a fixed carbon content of 83 to 93 percent and volatile matter of 10 to 15 percent, and is thus a smokeless fuel with a high heat value. It is free of soot and burns slowly, making it an excellent domestic fuel.

Anthracite is a hard, dense coal with the highest fixed carbon and lowest hydrocarbon content of all coals. It is characterized by a jet-black color, freedom from ash or moisture, and excellent coherence. It burns with a short blue flame and is an ideal domestic heating fuel.

Production Trends

Period of Decline 1918–1947

Although coal dominated in providing the nation's energy until World War I, the great growth in production of petroleum and natural gas initiated the beginning of stiff energy competition. Bituminous coal production fell from its peak of 579,386,000 tons in 1918 to 534,929,000 tons in 1929. During the Great Depression of the 1930s production declined drastically to 309,710,000 tons in 1932. During the remainder of the 1930s recovery was modest, reaching 394,855,000 tons in 1939.

During the World War II period, petroleum and natural gas were directed to military functions. As a consequence, bituminous coal production grew rapidly to a peak of 630,624,000 tons in 1947. After the war, the availability of low-cost petroleum and natural gas brought coal production down to a low of 402,977,000 tons in 1961.

Early Attempts to Stabilize Production

The history of the coal industry during the first half of this century was characterized by market instability, labor turmoil (the number of miners declined from a maximum of 704,793 in 1923 to 169,400 in 1960), uncertain profits, and difficult adaptation to governmental policies. The problems of the 1920s led to increasing problems and government concerns about the plight of a critical industry.

The first effort to address the problems was the 1931 creation of a private cartel among a group of coal producers in the Appalachian region. Appalachian Coal, Inc., was an attempt to establish a regional marketing cartel in order to maintain a minimum price with a prorationing system that would control output. Although Appalachian Coal was immediately challenged as a violation of the Sherman Antitrust Act, the Supreme Court declared the cartel legal. Shortly thereafter, more comprehensive federal legislation was passed by Congress to control the industry.

Nevertheless, the 1930s witnessed an extreme failure of the economic system. Instability resulted with boom and bust periods. As a response, there was institutional experimentation to create stability. As part of the New Deal, Congress created the National Recovery Administration (NRA), a mechanism designed to control the market by means of negotiations among producers and between producers and miners. The NRA set up "codes of fair practice" to prevent conflict and to stabilize production, distribution, and labor relations. The NRA attempt to manage the coal industry was an ambitious undertaking by a government dedicated to change. Regrettably it failed. Due to administrative, technical, and political reasons, the system was too unwieldy and vulnerable to abuse. In 1935 it was declared illegal by the Supreme Court.

In 1937 Congress passed the Bituminous Coal Act (Guffey Act) in order to establish an ambitious price/regulatory system that existed until 1943. The goal was to restore the coal industry to economic soundness. A minimum price was established for coals of different qualities. As the demand for coal rose during World War II, Congress abandoned the price-control experiment. Gradually it became apparent, in theoretical terms, that regulated minimum price increases, combined with the problems of excess capacity, made adjustments to reduce demand even more difficult. Because of the opportunities for fuel switching in energy, a price floor tends to enhance the competitiveness of alternative fuels, which created further decline of the industry. Unfortunately, few, if any, problems of the coal industry had been solved.

Cooperation and Stabilization, 1947–1970

In the 1950s and 1960s the coal industry adjusted to the changing economic environment. Prices stabilized and profits increased. Wages rose as mechanization of underground mines and strip mining brought improvements in worker productivity. The market for coal continued to shift to electric power utilities and away from residential, commercial, and industrial uses. Between 1960 and 1970 coal production rose from 187,420,000 tons to 602,932,000 tons. By 1975 electric utilities consumed two-thirds of the coal produced.

The key to the improved economic conditions was the emergence of a successful alliance between the United Mine Workers (UMW) and the large coal companies. While changes in mining technology and the coal market provided serious challenges, it was recognized by the UMW that if wages were to be increased, there had to be a decline in employment. The UMW also favored control of competition in the industry, which resulted in greater concentration of production with the elimination of the smaller, less efficient mining operations.

Between 1950 and 1970 there was essentially no control by the federal government. The Department of the Interior maintained a Bituminous Coal Advisory Council to advise the government. The coal industry, under the leadership of George Love, head of Consolidated Coal Company, established the Bituminous Coal Operation Association (BCOA) in 1950.

In the 20 years from 1950 to 1970, it was especially important that coal could compete with inexpensive oil and gas and consolidate its position as a preferred fuel in the electric utilities industry. Stable and low prices could best be achieved through rapid mechanization. In return for UMW cooperation in controlling labor costs the BCOA agreed to support a special welfare and retirement fund for miners and their families.

The large producers and consumers recognized the need to cooperate in regulating the coal market. As the sale of coal to utilities expanded, so did the proportion of long-term contracts. Utilities and steel companies also secured captive mines in order to prevent interruption of supply due to strikes. Operators and their customers built reserve stocks so that if strikes did occur, there would be no interruption of service. Finally, there was a series of mergers that transformed the structure of the coal industry.

In the period of 1950 to 1970 there is no question that the industry was significantly altered in favor of stability, assured profits, and overall production continuity. Labor peace prevailed. From the perspective of the total economy, most of these changes were constructive and beneficial. Consequently, a major U.S. industry became stable. The energy supply became more predictable, and communities dependent upon coal achieved greater stability.

Coal Industry, 1970–1990

A number of major events occurred in the 1970s that affected the U.S. coal industry. First, two federal acts were passed: the Federal

Mine Health and Safety Act of 1969, and the Clean Air Act of 1970 that limited the growth of the coal industry due to a new set of controls. In contrast, the oil crisis brought an awareness that the nation was not utilizing coal effectively in fulfilling energy requirements.

Coal and the 1970s Oil Crisis

The Arab oil embargo of 1973 and the dramatic rise in the price of oil triggered the public's rediscovery of coal. Suddenly, the United States began to be called "The Persian Gulf of Coal" and coal became the "great black hope." A new "age of coal" was envisioned. It was to be the key fuel in the development of energy self-sufficiency in the United States. Many felt that coal was to be the "transition" fuel until the inexhaustible energy resources, such as solar power and fusion, became available.

While President Nixon recognized the need to increase coal production, President Carter in April 1977 made coal's rebirth official when he introduced his National Energy Plan. He recognized the huge disparity between the United States' coal resources and coal's current use. Carter recognized that while coal comprised over 90 percent of the total U.S. energy resources, it provided only 18 percent of total energy consumption. President Carter's plan called for an 80 percent increase in coal production by 1985, from about 678 million tons in 1976 to over 1.2 billion tons. In addition, Carter proposed an expanded program of research and development in such areas as mining, burning, liquefying, and gasifying of coal, and also sought the conversion of scarce fuels to coal. Regrettably the implementation of the plan was delayed by other federal restraints and coal production in 1985 totaled only 878 million tons.

Clean Air Act

The basic purpose of the Clean Air Act is to control the pollution of the atmosphere. As the population of the nation has grown and as industry has expanded, the quantities of solid and gaseous contaminants have increased at a pronounced rate. Consequently, there is a growing concern that the pollution of the atmosphere over considerable areas has reached such a concentration that it interferes with the comfort, safety, and health of humans. Of the more than 170 million tons of air pollutants emitted into the atmosphere in the United States annually, about 60 percent is attributed to the automobile, 30 percent to the burning of coal and oil, and 10 percent to natural causes.

CLEAN AIR EMISSION STANDARDS   To control man-made pollutants the federal Environmental Protection Agency on December 1971, with revised standards in 1990, issued air pollution control regulations for the emission of sulfur dioxide (SO_2), oxides of nitrogen (NO_x), and particulates (fly ash). To the eastern United States' coal industry the standards for emission of sulfur dioxides from coal-fixed plants using high-sulfur coals were most important. Concern is greatest for sulfur dioxide because it requires a complex and costly cleaning process either before burning, or scrubbers must be used to remove it during the burning process.

A formula of SO_2 emissions devised by the EPA was expressed in terms of weight of SO_2, relative to the Btu present in the coal. The calculations are as follows:

Assuming coal with properties of 3 percent sulfur, 10,000 Btu's/lb, 20 million Btu's per ton (in order to understand the relationship of SO_2 to sulfur it must be remembered that the weight of sulfur is twice that of oxygen. Thus SO_2 formed in combustion is equal to twice the weight of sulfur contained in the coal).

The sulfur content of the coal is: $.03 \times 2{,}000$ lbs$=60$ lbs of sulfur emissions of SO_2; when coal is burned, 2×60 lbs per ton$=120$ lbs of SO_2 per ton; when related to heat content, 120 lbs SO_2 per ton/20 million Btu's per ton$=6$ lbs of SO_2 per million Btu's.

The EPA regulations classified coal into three categories of SO_2 content, each subject to a different level of SO_2 reduction. They are:

Category 1   All coal with an SO_2 content equal or less than 2 lbs/million Btu's must remove 70 percent of all SO_2.

Category 2   All coal with an SO_2 content between 2 lbs/million Btu's must reduce the SO_2 content so that final emissions do not exceed 0.6 lb/million Btu's. This requires a variable percent reduction of SO_2 by 70 to 90 percent.

Category 3   All coal with an SO_2 content between 6 lbs/million Btu's and 12 lbs/million Btu's must reduce SO_2 emissions by 90 percent. The maximum final emissions level, or ceiling, after 90 percent reduction is 1.2 lbs/million Btu's.

Under EPA regulations, the amount of sulfur in coal must be contained when burned. Sulfur occurs in several forms in coal. If it occurs in the pyritic form, such as iron sulfide (F_eS_2) or iron sulfate (F_eSO_4), it is amenable to partial removal (10 to 90 percent) by precombustion cleaning processes. In contrast, organic sulfur and residual pyritic sulfurs are released upon combustion in the form of SO_2.

In the United States most of the coals east of the Mississippi River are high in sulfur and most western coals in such states as Wyoming, Montana, Colorado, Utah, and Arizona and a few eastern coal seams, primarily in eastern Kentucky, are low-sulfur coals and can comply with the clean air standards without costly cleaning and emission controls. In essence the EPA regulations of the Clean Air Act created two regional markets for steam coal: high-sulfur versus low-sulfur coals.

In 1970 U.S. coal production was 602,932,000 tons. Six states (West Virginia, Kentucky, Pennsylvania, Illinois, Ohio, and Virginia) with 83.8 percent of that total dominated production. In 1979 coal production had increased to 776,299,000 tons, but the share of the coal production from these six states had fallen to 63.6 percent. By 1990 the eastern output was 53.6 percent out of a total production of 1,029,076,000 tons.

The growth of mining in western states has been spectacular. Wyoming, the leader, increased its coal production from 7,222,000 tons in 1970, to 70,204,000 in 1979, and 184,249,000 tons in 1990. In 1970 the western states of Wyoming, Montana, North Dakota, Colorado, and Arizona produced less than 5 percent of the nation's coal. By 1990 these states produced 26.1 percent of the total.

The increase in western coal is primarily due to the lack of sulfur. A large percentage of the coal has low Btu's since most of the coal of Wyoming and North Dakota is subbituminous. The greatest increase in Wyoming is due to the development of a number of factors. A number of coal seams that vary from 12 to 20 feet thick are located on or near the surface. Because the coal is consumed in the Midwest, unit trains of 150 to 200 freight cars move coal nonstop to the east at speeds of 75 to 90 miles per hour. These trains move tremendous quantities of coal great distances at very low cost.

The low-sulfur coals have thus experienced remarkable growth. In the high-sulfur coal areas of Appalachia and the Midwest the industry has experienced, at best, stability, and in most states decline. For example, in Pennsylvania coal production rose from 80,491,000 tons in 1970 to 87,069,000 in 1980 but has since declined to about 66,000,000 in the early 1990s. The coal industry has survived in the eastern United States due to: (1) utilization of coal in electric power plants, usually in areas of low population where high-sulfur coals can be burned after cleaning; (2) the production of metallurgical coking coal; and (3) the export market.

Federal Mine Health and Safety Act

Since the beginning of the coal mining industry technological progress has attempted to make it a less hazardous occupation. National standards were, however, not established until the passage of the Federal Mine Health and Safety Act in 1969. The purpose of this act was "to provide more effective means and measures for improving the working conditions and practices in the nation's coal or other mines in order to prevent death and other serious physical harm, and in order to prevent occupational diseases originating in such mines." The act established mandatory health and safety standards and required each operator of a coal mine and every miner to comply with the established standards. The act also had provisions to expand research and training programs aimed at preventing coal mine accidents and occupationally caused diseases in the industry.

In order to enforce the regulations, inspection of underground and surface mines is required on a regular basis. The purpose of the inspection is to: (1) obtain and disseminate information relating to health and safety conditions, (2) gather information on mandatory health and safety standards, (3) determine whether a danger exists in the mine, and (4) determine whether the mine operator is in compliance with the mandatory health and safety standards.

While the Clean Air Act altered the national pattern of coal production, the Federal Health and Safety Act, by influencing productivity, had an effect on the localization of both underground and surface mining. To implement the safety standards many workers had to be hired who had not mined coal. As a consequence, the productivity of the mine declined. Output per man per day declined about 20 percent. In addition it required additional funds to provide the safety features necessary to operate a mine. As a consequence, the small marginal mines were forced out of production. Only the larger mines were able to survive. Further, surface mining was encouraged because these mines required fewer workers to implement safety measures. The higher costs associated with the act encouraged technological advancements to improve productivity. The most significant change has been the development of the long wall mining procedure with the gradual abandonment of the traditional room and pillar systems. Thus this act has greatly influenced the growth of large, technically advanced mines and the decline of the small mine.

Future

Although total production is rising, the market is essentially limited to the production of electricity. Further, vast areas of the highest grade bituminous coal have limited output due to high sulfur content. In order to utilize bituminous coal effectively a major research program is required to develop the technology to utilize the high-sulfur coals. In order to utilize coal it must be not only environmentally acceptable but also cost competitive.

Anthracite

The northeastern Pennsylvania anthracite coal fields possess the only major anthracite deposits in the United States. In the nineteenth and early twentieth centuries the demand for anthracite was so great that a mining economy dominated the region.

Location

The surface area of the anthracite field is remarkably small, covering only 484 square miles in a 10-county area. The northern field of 176 square miles extends through Luzerne, Lackawanna, and small portions of Susquehanna and Wayne counties. The largest cities of the northern field are Scranton and Wilkes-Barre. The eastern middle field, centering on Luzerne County with extensions into Schuylkill, Carbon, and Columbia counties, has an area of only 33 square miles. Hazelton is the urban center of this field. The western middle field occupies 94 square miles in Northumberland, Columbia, and Schuylkill counties. The largest urban center is Shamokin. The southern field is the largest in area occupying 180 square miles. It extends northeast-southwest in Schuylkill, Carbon, Dauphin, and Lebanon counties.

Geology and Reserves

The Pennsylvania anthracite fields, lying in the folded and faulted Ridge and Valley Province, are geologically complex. The coal fields have been subjected to intense folding into synclines and anticlines in which faulting is common. The structured complexity increases from north to south, so that in the north the folds are open and symmetrical, while in the south there is tight folding with associated high-angle faults. The intense earth pressures applied over millions of years have reduced the gaseous content of the coal leaving only the carbon materials.

Over 200 coal seams have been mined in the anthracite fields, ranging from 1.1 to 19.3 feet in thickness. Most of the seams are under 5 feet in thickness. The original reserve of coal was estimated at 24.4 billion tons, of which about 9 billion tons have been mined or lost in the mining process. Of the 15.4 billion tons that remain, about 7 million tons could be recovered by modern mining technology. Of the remaining reserves, about 6.9 billion tons are underground and 100 million tons can be recovered by surface mining. The remaining reserves are, however, not evenly distributed. Mining was initially concentrated in the less geologically complex areas so that the reserves in the northern field are largely depleted. The southern field, however, has large reserves totaling about 9.5 billion tons with an additional 3.5 billion tons in the middle fields.

Production Trends

The anthracite mining industry has experienced two distinct production trends. The first period, extending from initial exploitation in the 18th century, was characterized by a steady growth in production to a peak of 100,445,000 tons in 1917. Since then production has declined steadily from 46,339,000 tons in 1950, to 9,248,800 tons in 1970, and 3,447,000 tons in 1990.

Period of Growth

The first recorded use of anthracite was in 1769 when Obadiah Gore used it in his blacksmith shop in Wilkes-Barre. The use of anthracite increased slowly because there was an abundance of firewood, it was difficult to ignite, and many problems had to be solved in transporting the fuel to the East Coast market.

The first major market for anthracite was in the smelting of iron ore. The anthracite furnace was larger than the charcoal furnace so it assumed a major role in supplying the growing U.S. iron market. In 1841 a dozen anthracite furnaces were in operation in eastern Pennsylvania and four in New Jersey. The Lehigh Valley was the center of the new industry. By 1860 more than a million tons of iron were produced in the anthracite furnaces. Anthracite was beginning its 20-year reign as the chief furnace fuel. By the late 1870s coke produced from bituminous coal assumed leadership and the industry shifted to western Pennsylvania.

As the demand for anthracite declined in iron ore smelting, home heating became the dominant market. As the forests were depleted in New England and the East Coast, a new domestic heating fuel was sought. Anthracite was the fuel nearest these expanding markets and it was considered a near-perfect heating fuel. Because of its high carbon content of 92 to 98 percent fixed carbon and 2 to 8 percent hydrocarbon, it possesses a large amount of heat (averaging 12,695 Btu/lb) and is nearly free of ash and sulfur. It burns nearly smokeless.

In order to develop the anthracite market the first prerequisite was the development of a transportation system. This was first achieved by establishment of a canal system. Between 1827 and 1846, 643 miles of waterways were canalized. Canals greatly reduced shipping costs, giving anthracite a competitive advantage over other fuels on the East Coast.

The canals were replaced by the railroads. Railroad trackage increased from about 1,000 miles in 1860 to 2,290 in 1873. In 1900 there were 10 railroads hauling anthracite exclusively, providing a dense network in the coal fields with each serving a specific East Coast market.

Period of Decline

Many reasons have been given for the decline of the anthracite industry. Of these the depletion of reserves has often been suggested as the most-likely cause, since anthracite is a nonrenewable resource and billions of tons have been mined. This occurs when a mine is depleted of its coal deposit and a specific settlement, where mining is the only source of employment, may not have an economic base. However, with billions of tons of coal reserves remaining, the drastic decline in production cannot be found in coal depletion but other factors must be sought.

It has also been suggested that because anthracite mining is technically difficult due to geological conditions, mining costs have become prohibitive. Because this is an old mining area and since many of the more accessible underground seams are depleted, and stripping operations have mined the beds nearest the surface, there is little doubt that mining is becoming more difficult. A general rising curve of production per man per day from 2.83 tons in 1950 to about 7.20 in the 1980s indicates that technical problems are being solved. Difficult engineering problems do not appear to provide a satisfactory explanation for the production decline.

A more plausible explanation for the decline in the use of anthracite appears in the increasing availability and consumption of petroleum and natural gas. In 1920 anthracite supplied more than 95 percent of the home heating needs of the area east and north of the Pennsylvania anthracite fields. As competition from petroleum and natural gas increased, anthracite's share of the market declined sharply. By the early 1950s fuel oil and natural gas supplied about 66 percent, anthracite 32 percent, and coke and briquettes about 2 percent of the home heating fuel needs of this area. By the 1980s the demand for anthracite as a home heating fuel had dropped to insignificance.

The question immediately arises as to whether this loss of demand in the traditional anthracite market area was due to a price differential between competing fuels. A study by the U.S. Bureau of Labor Statistics in 1952 revealed that the price of anthracite for many grades was considerably lower in eastern Pennsylvania than that of fuel oil or natural gas for an equal amount of heat. In New York, for example, the heat equivalent of fuel oil from 1 ton of pea grade anthracite was $23.49, while the cost of pea grade anthracite was $19.90. Natural gas had an even greater cost disadvantage over anthracite. For the equivalent of 1 ton of pea coal, natural gas in New York cost $58.74 and in Baltimore $37.73. When cost differentials existed against anthracite they were very slight. These cost differentials have changed little over time, indicating that cost has not been a major factor when selecting a residential heating fuel.

Two principal factors appear to explain the change in fuel used to heat homes. First is the convenience in the use of fuel oil and natural gas; and, second, are the psychological determinants that evolved after 1920. The trend toward the use of more and more laborsaving devices in the household was of the greatest importance. The American home could be heated through no more effort than the delicate adjustment of a thermostat. In addition, tons of anthracite did not have to be stored in a basement area, in a coal shed, or an unsightly open pile in the backyard of a home. Further, anthracite had to be shoveled into a furnace or a stoker and, after burning, its ashes had to be removed. Long after the decline had begun, the anthracite industry, largely through research conducted through the Anthracite Institute, did attempt to develop and promote the automatic stoker. These efforts were largely unsuccessful. An additional factor that discouraged the use of anthracite in home heating was the difficulty of igniting anthracite in the furnace. Because anthracite has little hydrocarbon gases, it requires a high ignition temperature.

The psychological factors, though more elusive, are no less real. The use of anthracite was no longer considered fashionable. It had become an antiquated fuel. When economic factors were of little importance, other considerations began to play a crucial role. In the mode of the times, "keeping up with the Joneses" required a "modern" fuel to heat the home.

Besides these major factors a number of lesser factors contributed to the decline of anthracite. Several labor strikes, particularly one lasting 170 days in the winter of 1925–1926, gave the impression to the general public that anthracite might be in short supply, or even unavailable, during the cold winter season. Many anthracite users turned to the use of fuel oil or natural gas. To implement this change, the coal furnace was discarded and a new furnace purchased, or the old furnace was converted at considerable cost to use the new fuel. While anthracite was never in short supply due to company stockpiling, fear by the general public created an awareness that caused a shift to other energy sources.

In mining anthracite there is frequently a large amount of slate and other rock in the original tonnage. The breakers removed some of this rock, but a large quantity remained. When the coal-producing companies had little or no competition from other fuels, the general public was forced to purchase coal that contained hundreds of pounds of slate in each ton. The coal companies for many years did not recognize the seriousness of fuel oil and natural gas competition and continued to exercise poor quality control in the industry. The general public gradually lost confidence in the industry's ability to provide a quality product. This sapped the marketing potential of anthracite.

As the market for anthracite as a home heating fuel declined a number of smaller, auxiliary markets were affected. Originally, essentially all of the coal was marketed by rail transportation. The anthracite railroads used anthracite as a fuel. Because their freight consisted primarily of coal, railroad consumption declined as the domestic market vanished. To aggravate this situation, trucks gradually replaced the railroads in transporting coal. With the abandonment of deep mines, consumption of coal declined at the mines also.

Anthracite has not been able to expand its market beyond home heating. A market in manufacturing and public utilities has not evolved. In fact, anthracite use as a fuel to produce cement has disappeared. In electric power generation, anthracite requires entirely different equipment than that used in bituminous fired plants. Higher ignition temperatures and longer burning time necessitates a much larger boiler to provide the longer flame path. The boiler sizes for anthracite must be about twice that of equivalent capacity bituminous boilers. Because anthracite is much harder than bituminous coal, grinding is more costly. While there is little sulfur emissions, the control of particulates (fly ash) requires an expensive bag filter system. Thus, the capital expenditure for an anthracite power facility is much greater than for a plant using bituminous coal or petroleum.

Future

Although billions of tons of anthracite remain located in one of the areas of highest energy consumption in the United States, anthracite has played no role in modern-day attempts to develop a U.S. energy policy. Anthracite is not recognized as one of the nation's energy resources. State and federal legislation to encourage development of this energy source has been lacking. No evidence exists today that this policy will be changed in the immediate future.

Nuclear Energy

The production of electricity from atomic materials began at a reactor at Shippingport, Pennsylvania, in 1957. The taming of the atom provided great hope that atomic energy would ultimately replace fossil fuels in the production of electricity. Atomic energy was heralded as a cheap, clean, efficient and safe energy source for the world.

Growth of Nuclear Power

In the 1960s nuclear power grew slowly. By 1965 there were six generating units in the United States. This number grew to 18 in 1970. These units produced 22 billion kilowatt-hours (KWh), which was about 1.4 percent of the total electric utility generation. The 1970s witnessed the major growth of atomic energy with the number of reactors rising from 18 to 70 in 1980, and the production of electricity rising to 251 billion KWh.

Although there have been hundreds of small accidents in the production of electricity from nuclear fuels, the first major accident in the United States occurred at Three Mile Island, near Harrisburg, Pennsylvania, on March 28, 1979. Since that time there has been a significant slowdown in the development of nuclear electric power. This has been due to the public fear of radiation from accidents at a nuclear plant plus the sharp decrease in the projected rate of increase in electric power demand and the rising cost of constructing new nuclear plants.

Between 1970 and 1990 the number of operable generating units grew to 111 producing 576.8 billion KWh, or about 20.6 percent of the total electric generation in the United States. Most significant, however, has been the increase in the number of plants producing over a million kilowatts annually. In 1982 there were 12 such plants but by 1992 this number had increased to 49.

The nuclear power industry is widely scattered in the United States. The South Atlantic states with 26 plants has the largest number followed by East North Central with 23 plants. Illinois with 13 plants has the highest number of units, followed by Pennsylvania with 8 and South Carolina with 7.

Nuclear Fuel Cycle

There are numerous processes in the nuclear fuel cycle. The main steps are:

Mining of the ore

Milling

Conversion

Enrichment

Fuel Fabrication

Energy Production (nuclear energy)

Fuel Reprocessing

Waste Management

At each of these steps there is some danger of exposing the public to harmful radiation. The greatest dangers stem from the possibility of a reactor accident during energy production and the problem of nuclear waste disposal after energy production takes place.

Mining

The danger of radiation contamination in mining radioactive materials was long ignored. Prior to 1930, uranium ores were mined for radium in Arizona, Utah, Nevada, Colorado, and New Mexico. Beginning in 1946 large-scale uranium mining began, first for atomic weapons production and later for nuclear electricity. Intensive mining continued through 1968, when existence of adequate stockpiles of uranium led to a reduction of mining activity. During that period more than 6,000 miners were exposed to radiation.

Radon measurements were first made in uranium mines in 1949. The surveys were, however, somewhat spotty and largely carried out to collect information rather than to control radioactivity in mining operations. Over the years, evidence accumulated that radioactivity was high in uranium mining operations and could be harmful to the miners' health. In 1957 the U.S. Public Health Service predicted a significant mortality rate from lung cancer among miners. Finally, in 1961 major radon control programs were initiated in underground mines. Advances in ventilation technology have continually improved the situation. Although the risk of acquiring lung cancer has not been eliminated for uranium miners, it has been greatly reduced.

Milling

In the milling of uranium ore more than 99 percent is waste, so near the mill sites huge piles of tailings accumulate. The extraction of a few pounds of uranium to fuel a single 1,000-megawatt light water reactor for a year requires the generation of at least 106 tons of tailings. For many years the danger from radioactivity was not recognized. Because of the sheer volume, the Environmental Protection Agency (EPA) believes that mill tailings may present the greatest environmental impact of all waste forms.

For many years the fine-ground tailings were used for many purposes by the public. For example, tailings were used at construction sites of homes and malls in Grand Junction, Colorado, for more than 15 years. In 1966 Colorado public health officials found abnormally high readings of gamma radiation and radon in certain buildings. Further investigation revealed that over 3,000 homes and buildings had tailings under their foundation.

The solution to the mill tailings problem is not an easy one. The best solution is the complete removal of the tailings. Because of the massive amount of material and because it is scattered over wide areas of the southwestern United States, the cost could be as much as $100 million. Several measures can be taken to minimize the radon hazard from tailings. The simplest procedure is to flatten the piles and cover them with earth. One objection to this technique is, for example, that the half-life of radium 226 is 1,630 years. Thus, radiation is likely to exist longer than the protection of a thin earth covering. The best procedure would be to return the tailings to the mill and recover the radium content. Because of the small amount of radium involved, this procedure is costly. The most sophisticated solution would be to chemically recover the radium, incorporate it with the reactor fuel, and destroy it by transmutation and fission.

Enrichment

The next step in the nuclear fuel cycle is the enrichment of the hexafluoride (UF_6), which becomes a gas at conditions near room temperature and pressure. Enrichment is accomplished by a gaseous diffusion process. In this process large quantities of heat, water, and electricity are required. The large amount of cooling water required is the major environmental impact in the enrichment process. Cooling towers are required to reduce the temperature of the water. Other environmental impacts are minimal. The radiation dose to an individual living at the plant boundary would be only 2 mrem/year (the rem is defined as a dose of radiation required to produce the biological effect of one roentgen of gammaradiation. A roentgen [R] is a measure of the ability of gamma rays to produce ionization).

Fuel Fabrication

The final step before the fuel is used in the nuclear reactor takes place when the enriched UF_6 is converted into uranium oxide (UO_2). The UO_2 is formed into small ceramic pellets and enclosed in a thin tube made of a suitable material, usually zircalloy, to form fuel rods. The fuel rods are then transported to the power facility in bundles called fuel assemblies. The Environmental Protection Agency has indicated that this process has little or no impact on the environment.

Energy Production

One of the most controversial issues in the development of atomic energy is the fission of atoms in the nuclear reactor. There has always been apprehension about the safety of nuclear reactors, and this controversy has grown with the Three Mile Island and Chernobyl accidents.

An atom is the smallest particle into which an element can be divided chemically. It consists of positively charged particles called protons and an equal number of neutral particles called neutrons. The number of protons in the nucleus determines the chemical properties of the element. Uranium has 92 protons. Atoms of the same element, however, may have a different number of neutrons. These are known as isotopes. For example, uranium 238 (U238) has 146 neutrons and uranium 235 (U235) has 143 neutrons. This means that the chemical properties of these two isotopes are identical but the nuclear properties are fundamentally different.

Nuclear fission occurs when certain elements are split by neutron bombardment. Of all elements, uranium is the only fissionable material that occurs in large quantities in nature. Thorium is also fissionable, but occurs only in small quantities in nature. The uranium atom has a number of isotopes but can be converted into uranium 235 when placed inside a reactor.

When uranium 235 absorbs an additional neutron, it will likely split, or fission, into two smaller nuclei, resulting in the release of a large quantity of energy and the emission of two or three neutrons. The emitted neutrons are now available for absorption by other fissionable nuclei. In this manner a fission reaction is made self-sustaining if the neutron emitted by one reaction triggers a subsequent fission reaction. Not all neutrons emitted in the fission process initiate subsequent fission. Some neutrons escape from the area of reaction, some are absorbed, and some are captured by heavy nuclei without producing fission. The rate of fission is controlled with moderators—substances used to slow down neutrons in a nuclear reactor. If the emitted neutrons were not controlled, the rate of fissioning and consequent heat output would increase exponentially, resulting in the ultimate explosion of the reactor.

Fuel Reprocessing

The reactor's fuel cells must be replaced when they are spent. Because the spent rods are highly radioactive, the normal procedure is to store them about a year underwater, during which time the most intense short-lived and intermediate half-life radioactive products have reduced their radioactivity. The rods are then transported to a reprocessing plant to separate residual uranium and plutonium. The recovered uranium, which could still contain around one percent U235, can be converted into uranium hexafluoride (UF_6) for subsequent enrichment. The recovered plutonium can be converted into plutonium dioxide for subsequent use in mixed oxide fuel elements that are a blend of uranium and plutonium dioxides. The fission products containing the radioactive wastes must be treated and disposed of.

Uranium enrichment and reprocessing of spent fuel are considered the most sensitive elements of the fuel cycle from the nonproliferation point of view. Although initially the recovered plutonium is not weapon grade, it is technically possible to upgrade it in quality for the production of weapons. Transfer of technology and international cooperation in this area has essentially been lacking.

The Environmental Protection Agency has attempted to predict the health affects at a reprocessing plant. It is estimated that a plant with a capacity of 5 metric tons per day serving 45 nuclear plants of 1,000-megawatt size would emit long-lived radioactive gases such as krypton 85 and tritium at a rate that would produce about 2.5 health effects (cancer due to radiation) annually in the United States, and about 100 health effects annually worldwide.

Nuclear Waste

The final step in the nuclear fuel cycle is management and disposal of radioactive waste, some of which may have a radioactive half-life of tens of thousands of years. Some forms of waste are generated at nearly every stage of the nuclear fuel cycle, from the tailings at the mill site to the high-level fuel reprocessing operations. The problem has attracted much attention, and public fears have become so pervasive that the entire nuclear industry has been threatened.

In the process of producing electricity, for every metric ton of nuclear fuel in an initial load, 24 kilograms (kg) of uranium 238 and 25 kg of uranium 235 are consumed in a three-year period. In the fission process the enriched uranium 235 is reduced from 3.3 percent to 0.8 percent of the total. The consumption of 1 ton of nuclear fuel produces 800 million kilowatt-hours of electrical energy.

In the process of producing electricity, 35 kg of radioactive waste products are produced in the form of isotopes in the following amounts: 8.9 kg of plutonium, 4.6 kg of uranium 236, 0.5 kg of neptunium 237, 0.12 kg of americium 243, and 0.04 kg of cirium 244. Because only 25 kg of uranium 235 are consumed, and a fifth of that amount converted into uranium 236 and neptunium 237, it is calculated that only 60 percent of the energy produced comes from uranium 235.

Although the nuclear power industry is more than 35 years old in the United States, all nuclear waste has been placed in temporary depositories. These depositories contain more than 98.8 percent of the nongaseous fission products produced in the reactor. At present, these liquid wastes are usually concentrated by evaporation and stored as an aqueous nitric acid solution in high-integrity, stainless steel tanks. There is now general agreement in the nuclear industry that wastes should be solidified and then placed in a permanent depository. It is now estimated that the United States has more than 16,000 tons of nuclear wastes.

Numerous studies of the types of permanent disposal of nuclear waste have been conducted. One of the major problems when choosing and identifying suitable sites is the vocal objection of the general public. Although there is universal recognition that the waste problem must be disposed of, there is also the "not in my backyard" viewpoint. The problem is thus not only a technical one but also one with emotional and political ramifications.

In the search for a place to store high-level radioactive wastes, three sites were originally selected. These were Hanford, Washington; Hereford, Texas; and the Nevada Atomic Test Site, Yucca Mountain, in the desert of southern Nevada. Although there was opposition from all of the states, the Nevada site was chosen.

In the Yucca Mountain area geologists and engineers are beginning the task of preparing an underground depository to isolate high-level radioactive waste. The estimated cost of the project is $2 billion and the completion date is 2003. Teams of geologists and engineers are attempting to predict whether the area provides a safe depository for high-level waste. Though some of the waste that would be deposited could be hazardous for millions of years, predictions are being limited to 10,000 years. Beyond this time span, most questions of safety are simply unanswerable.

There has been conflict between the geologists and engineers on the project. Whereas engineers complain that the geologists' predictions reach beyond the limits of their expertise, geologists contend that the engineers sacrifice scientific rigor for adherence to procedures used to ensure only satisfactory present-day engineering practices. At the Yucca Mountain site geologists have clashed with engineers who insist that all work proceed according to a system called "quality assurance." Under this system, engineers plan and execute a job, then check to see if the plan was followed precisely in order to guarantee satisfactory work. Geologists, in speculating what will happen in the distant future, cannot follow such a structured procedure.

Mankind is faced with one of its most difficult problems. Completely rational decisions may not be possible. Nevertheless, there can be no question, a permanent depository for high-level nuclear wastes is fundamental to the welfare of all people.

Future

The production of electricity from atomic energy has developed in most industrial nations and is advancing in a number of Third World countries. The importance of nuclear power varies greatly from country to country. France, with the greatest dependence, secures about two-thirds of its electricity from nuclear power.

In the United States the future of nuclear power to supply electricity received its first major setback in 1979 with the Three Mile Island nuclear accident. The issues were compounded by the Chernobyl accident of April 26, 1986. Out of these incidences a moral, political, and economic controversy has evolved that threatens the future of nuclear power. The nuclear industry has had little development in the United States. There have been no new orders for plants in the United States since 1973. Most significant, there has been the cancellation of more than 116 plants, representing investments estimated at more than $20 billion while only 109 plants have been completed. Due to increasing governmental regulation the construction time doubled, and the cost of new plants soared six- to eight-fold over original estimates. The plants of the nuclear power industry are now 20 or more years old and may need to be modernized.

There have also been delaying tactics to placing a number of completed plants in operation. In 1987 three governors—Michael Dukakis of Massachusetts, Mario Cuomo of New York, and Richard Celeste of Ohio—took stands against placing controversial nuclear plants in operation. Citing health and safety concerns, the three governors refused to certify emergency evacuation plans for, respectively, Seabrook in New Hampshire, Shoreham in New York, and Perry and David Besse in Ohio.

The major accidents plus minor shutdowns of nuclear facilities brought to the forefront questions about the safety of nuclear power. In addition, the Three Mile Island cleanup was not completed in 1993, adding to public apprehension. Never before has a simple event had such a long-term influence on the development of a basic resource—energy—vital to the U.S. economy.

In order to attain greater safety, the nuclear industry has improved its efficiency. As Scott Peters of the nuclear industry's U.S. Council for Energy Awareness states, "We're performing better than we ever have. If you're performing better, you're performing safer." Harold Denton of the Nuclear Regulatory Commission (NRC) agrees that plants are definitely safer today than they were before the Three Mile Island accident. The confidence of the general public in plant safety, however, is not assured by the continued number of incidents at nuclear plants. Plants must report all "potentially significant safety accidents" ranging from a tool improperly left in an electric cabinet to a flawed reactor. In the four-year period from 1984 through 1987, 11,400 incidents were reported of which 8,400 were personal errors. Another measure of efficiency is the number of automatic shutdowns that occur at a plant during a year's operation. In 1981 there were 487 automatic shutdowns, or an average of 5.24 per plant. By 1987 this number had been reduced to 342 or only 3.25 shutdowns per plant. The number of automatic shutdowns continues to decline, but are still too high.

Public support for nuclear power remains quite low. These present-day attitudes are a response to the continuing pattern of technical problems and misinformation. As the most sophisticated and potentially dangerous technology ever harnessed to provide a basic need, nuclear power requires an extraordinary faith by ordinary citizens.

In spite of public opposition, it is now recognized that some other form of energy must replace fossil fuels—particularly petroleum and natural gas—as generators of electricity. Many world leaders recognize the need for nuclear power in the future. West German Chancellor Helmut Kohl stated, "Abandoning nuclear power would spell the end of the Federal Republic as an industrialized nation." Energy Secretary Peter Wolber of the United Kingdom said, "If we care about the standard of living of generations yet to come, we must meet the challenge of the nuclear age and not retreat into the irresponsible course of leaving our children and grandchildren a world in deep and probably irreversible decline."

Many fundamental questions concerning the future of nuclear power remain unresolved at the present time. Some of these are: Can public confidence in nuclear power be restored in the foreseeable future? Will the slow growth of nuclear power in the 1980s and 1990s limit technological advances to a safer industry? Can the moral and political aspects of nuclear development be reconciled with the science and technological ramifications in the future? Finally, and possibly most important, in a world economy based on vast energy consumption, can another source of energy to produce electricity be developed to replace nuclear energy as the nonrenewable energy supplies dwindle in the next century?

Renewable Energy

Renewable energy plays an insignificant role in supplying the energy of the nation. Of the renewable resources, hydropower is the most important. Only recently during the energy crisis of the 1970s has attention been directed to the potential of other sources of renewable energy. At best, development has been modest and, with the present availability of traditional fossil fuels, the development of renewable energy in the immediate future will be minimal.

Classification

The Department of Energy has classified renewable energy into three major groups and eight individual categories.

1. Geothermal (heating and cooling) applications
    a. Heating and cooling of buildings, including hot water heating
    b. Agricultural and industrial process heating
2. Fuel from biomass
    a. Plant matter, including wood and waste
3. Solar electric
    a. Solar thermal electric units such as the "power tower"
    b. Photovaltaics—solar cells
    c. Wind/windmills
    d. Ocean thermal electric
    e. Hydropower/hydroelectric dams

Evolution of Renewable Energy

The Industrial Revolution was founded on the availability of cheap, abundant fossil fuels. The vast resources of fossil fuels made their use nearly invisible and waste painless. As a consequence, renewable energy resources have been little utilized providing only a tiny fraction of the nation's energy requirements. In 1952 the prestigious U.S. President's Materials Policy Commission observed, "Efforts to date to harness solar energy are infinitesimal. It is time for aggressive research in the whole field of solar energy—an effort in which the United States could make an immense contribution to the welfare of the free world." This recommendation was ignored completely by not only the government but energy corporations.

In 1955 when the International Solar Energy Society was established only 12 people were present and only two solar equipment companies existed in the world. During the following decade, fewer than 30 homes using solar energy were built in the United States. When photovoltaic cells were invented in 1954, they were considered only applicable for space applications. A number of wind experiments in the Great Plains and in New England in the 1950s failed as rural electrification programs developed. Hydroelectric power developments were the one exception to the use of renewable energy; large power dams were built in the western United States and the Tennessee Valley Authority developed the hydroelectric potential in the South.

The low cost and huge reserves of fossil fuels fostered the belief that there would be no fossil fuel shortage and discouraged development of renewable resources. The quadrupling of the price of oil in the 1970s and the shortage of gasoline had a devastating impact on the general public.

An initial response to the energy crisis of the 1970s was a consideration of national renewable energy resources. Immediately following the 1973 crisis, the wood product industry intensified the use of wood stoves. Energy conservation was practiced for the first time. The general public assumed that the era of low-cost, available energy had ended.

The development of renewable energy resources gained support as a measure to achieve a greater energy independence at a national level. The renewable resources could increase the flexibility of the nation's energy supply, while decreasing the adverse effect of international imports.

The federal government's response to the development of renewable energy accelerated during the 1970s. In 1974 when the Energy Research and Development Administration was created, renewable energy became one of its six programs.

However, renewable energy programs were funded at a minimal level receiving $14.8 million out of a total budget of $1.034 billion. Nevertheless, it marked the initial federal efforts. Most of the early federal activity involved research in three interrelated areas: technology development, commercialization barriers and impacts, and small-scale demonstration programs.

Until 1977 programs in utilizing renewable resources were, at best, moderate. The energy situation of the nation had deteriorated. Domestic oil and natural gas production had declined, while between 1973 and 1977 oil imports had increased by more than 30 percent. With the implementation of the Clean Air Act coal production in high-sulfur coals of the eastern United States had declined. Natural gas shortages forced school and factory closures in the upper Midwest and dramatized the effect of energy problems.

By 1977 the initial studies on the technical feasibility of utilizing renewable energy had been completed. President Carter placed a high priority on conservation and development of renewable energy programs. With the establishment of the Solar Energy Research Institute (SERI) a goal was set to use solar energy in 2.5 million homes by 1985. By 1978 three initiatives had evolved:

1. The energy tax act provided federal credits to help reduce the initial cost barrier. Credits up to $2,000 were provided to homeowners who purchased an active solar or wind energy system for use in their principal residence. Ten to 13 percent tax credits were provided

for business installations. The federal excise tax was canceled for gasoline that contained 10 percent ethanol alcohol content.

2. The Public Utility Regulatory Policies Act (PURPA) requires utilities to interconnect with independent producers and purchase their electricity at prices equal to the utility's "full avoided cost." At a minimum, avoided costs equal the value of the fuel not bound by the utility as a result of the independent producer. At a maximum, avoided costs include a "capacity credit" based on the utility's reduced need to construct additional generating plants.

3. The Natural Gas Policy Act revised price controls that had suppressed natural gas prices. The price of newly discovered gas was allowed to increase gradually until complete decontrol in 1985.

From 1977 through 1981 renewable energy achieved real growth and wide expectations. To illustrate, the Federal Energy Regulatory Power commission received many requests for permits to develop small sites across the nation. Wood energy use increased by at least two-thirds. By 1981 nearly 200,000 residential solar heating units had been installed, generating over $600 million dollars in sales. Almost all manufacturers of prefabricated housing offered passive solar options. By 1981 it was estimated that total renewable energy production totaled 5.5 quads (one quad equals one quadrillion Btu's). The expanding government programs, technological improvements, and the tripling of OPEC prices in 1979 provided the basis for the great growth of renewable resources.

After 1980 the emphasis changed in the utilization of renewable resources. Energy conservation became the key to energy utilization. Consequently the demands for traditional fuels declined. Most significant were government policy changes. The Reagan administration opposed or greatly reduced the scale of most renewable energy programs. Congress authorized $525 million for renewable energy development for 1981–1983, but the Reagan administration reduced the size of the appropriation by 95 percent and then impounded the remaining funds. In addition, the Reagan administration opposed extending the important tax incentives enacted in the early 1980s. The lack of these incentives severely affected market effectiveness. As a response to these and other measures, programs in the use of renewable resources have been minimal since 1980.

Hydropower

Hydropower is the principal use of energy's renewable resources. Waterpower has been harnessed since the earliest civilization to perform tasks such as the grinding of grain. In modern times the first turbine was attached to a generator in Wisconsin in 1882. Most of today's hydroelectric generating stations are fed by a dam that directs water down to a turbine. Obviously the technical knowledge for construction of hydropower dams has been intensely developed.

In recent years hydropower has produced 12 to 14 percent of U.S. electricity output. In 1988 there were 2,029 hydroelectric power plants in the United States that ranged in size from 1 megawatt to the 6,180-megawatt Grand Coulee Dam in Washington. Hydropower thus produces more electricity than any other renewable source, while contributing no carbon dioxide to the atmosphere.

The development of hydropower has reduced the cost of electricity in the United States by providing the most efficient, most reliable, lowest cost source of electricity. Conversion efficiency generally ranges from 83 to 93 percent, and dams generate power an average of 95 percent of the time. The low cost is primarily due to the construction of large projects in the 1930s and 1950s when costs were low and, in addition, were subsidized by federal hydropower projects. About 46 percent of the nation's capacity is federally owned. Costs are further reduced by the 50- to 70-year lifespan typical of large facilities.

Interest in the development of small waterpower sites grew rapidly in the 1990s. Waterpower appeared to be a long-term renewable energy replacement for declining oil production. The independent developers rushed to claim potential sites in response to the passage of the Public Utility Regulatory Policies Act in which utilities were required to purchase electricity produced by small producers. Between 1978 and 1984 over 5,000 permits were requested to conduct feasibility studies at new and existing sites. Fewer than 3 percent of the 67,000 existing U.S. dams currently produce electricity. Most were built for flood control, but many could be retrofitted to generate electricity. The U.S. Department of Energy estimates that up to 10,000 megawatts of potential small-scale hydropower could be developed. In New England it is estimated that more than 1,750 old or idle small dams could be harnessed to produce more than 1,000 megawatts of electricity annually. The development of small hydro sites subsided after about 1985 and growth of hydropower has been modest in recent years. The U.S. hydroelectric capacity is presently 88,800 megawatts.

The Federal Energy Regulatory Commission (FERC) estimates that the total potential capacity of conventional hydropower dams, both large and small, is 147,000 megawatts, nearly double the currently developed capacity. A recent energy survey by the Solar Energy Research Institute indicated that the hydro potential could be increased to 5.1 quads. In addition, Canada could play a major role in supplying electricity to the United States. Canada has huge reserves of undeveloped hydropower resources and is presently selling electricity to New England.

The future development of hydropower in the United States depends on overcoming a number of environmental, regulatory, and political obstacles. A number of federal laws restrict development of hydropower that would destroy the natural character of the region.

Solar Energy

Solar energy is produced from solar radiation. It has numerous applications such as producing heat for buildings, and the creation of electrical power. The nation's solar energy is immense. On an average day the United States receives about 130 quads, or more than 650 times the nation's total daily energy consumption. Solar energy is greatest in the Southwest where there is the least amount of cloud covering. However, even in the humid, cloudy east there is a substantial amount of solar energy.

In solar thermal systems, solar collectors with mirrored surfaces concentrate light onto a receiver, which heats a liquid, driving a turbine generator to produce electricity. The two categories of systems are central receivers and distributed receivers. The central system consists of one receiver perched atop a large tower surrounded by a field of mirrors called heliostats. The heat is then focused on a single surface that may reach a temperature of 1280°F.

In the distributed system the reflectors, mirrors, and receivers are contained in a single unit, a disk, or a parabolic trough.

The parabolic trough focuses the heat on a tube-shaped receiver with temperatures up to 750°F. Disk systems have temperatures up to 3,600°F or hotter.

With the energy crisis of the 1970s a major endeavor to utilize solar energy began. Solar heating dominated these efforts. Heating applications focused almost exclusively on mechanically operated, active space heating. These systems generally involved a rectangular, glass-covered collector, often roof-mounted with an orientation to the south. These systems are called "low temperature" collectors since the temperatures are generally below 110°F. Normally, hot water systems supply 60 to 75 percent of the annual heat made in the Southwest and 50 to 60 percent elsewhere. Thus solar energy systems must normally be supplemented by a conventional heating system in order to provide heat during a period of cloudiness.

In the early 1970s the initial solar collectors from residential heating were plagued by equipment failure and installation problems. By the early 1980s these problems were solved. Installations have been concentrated in Hawaii, California, Florida, and in the high insolation areas of the Southwest. From 5 to 18 percent of the homes in these areas have solar collectors. More than 1 million active collectors are now found in the United States. The growth of solar heating has declined since the mid-1980s due to two factors: the price of heating oil has stabilized and tax credits, which reduced initial costs by 75 percent from federal and local tax rebates, were removed. As a result many solar collector manufacturers went out of business.

Although major advances in technology have been made, many systems are still in research and development stages. The low sunlight-to-electricity conversion efficiencies, large equipment costs, and the intermediate nature of the output prevents solar systems from competing with conventional electric production. Land requirements with access to sunlight are frequently cited as an additional constraint to solar electric development. However, economic factors rather than space are the greatest new restraint on development.

Besides the home residential solar systems, large solar thermal electric generators progressed from laboratory experiments to demonstration-scale generating plants in the early 1980s. Solar One, a solar power tower came on line in 1982 near Barstow, California. During the summer, it is able to generate 10 megawatts of electricity for 8 homes a day from a 72-acre field of heliostats (slightly parabolic mirrors that reflect and concentrate solar energy into a central stainless steel receiver). The technology for this development is extremely complex. For example, each of the 1,818 heliostats has to reflect the sun's rays precisely onto a central receiver. The tracking path for each heliostat varies for each day of the year. Similarly, the levels in the central receiver must be able to withstand the equivalent of 300 sun's of energy impact and then zero impact within the 10 seconds it takes a sharp-edged cloud to obstruct the heliostat field. The project costs $141 million or approximately $14,100 per kilowatt of generating power. This is more than six times the typical cost of a kilowatt generated in a nuclear plant. While costs can be reduced, the cost is still prohibitive for solar energy electric development in the immediate future.

Photovoltaic Cells

Photovoltaic cells are the most technically advanced of the renewable energy techniques. These cells are specialized semiconductors that convert sunlight directly into electricity using advanced electronic processes. They have the potential of being able to tap energy directly from the sun, employing inexpensive and abundant raw materials such as silicon.

The theory for the construction of photovoltaic cells was first presented by Albert Einstein in his classic 1905 paper. In cell production, pure silicon is selectively "contaminated" with different gaseous impurities at high temperatures. These impurities define electrical boundaries between the elements of the cells that are connected by a thin layer of aluminum on the circuit. The resulting cells are photovoltaic, that is, they generate electricity. The early cells were manufactured under tight controls and costly procedures so they were extremely expensive—as much as $1,000 per watt, which was more than 1,000 times the cost of producing electricity by conventional means.

The early photovoltaic cells were used to provide electricity for orbiting space satellites. In these endeavors, cost was not a consideration, but technical performance and reliability were key considerations. However, with development, the cost of the photovoltaic cell has experienced a dramatic decrease and is now under $10 per peak watt. While a major cost reduction has occurred because of the ability to produce crystal silicon cheaply, it is still too high to compete with conventional fuels. In order to reach a competitive level, the cell must cost no more than $1 per peak watt. Most scientists do not believe this cost level will be achieved soon.

When the cost barrier is crossed, the potential for photovoltaic cells is very large. The cells then will be utilized to produce electricity and solar heat buildings. In order to achieve this goal there needs to be major research and development programs. This does not appear likely in the next decade.

Geothermal Energy

Geothermal energy has been used in small amounts for centuries. Hot springs have flowed naturally from the earth providing heat for cooking and home heating. The power from natural thermal steam was first used to produce electricity in 1904 in Italy.

Because geothermal energy comes directly from the earth its temperature varies greatly. At moderate temperatures (less than 350°F) it is normally used in direct heating applications, such as residential heating or industrial processing. In most of these uses geothermal energy is less costly than fossil fuels. It is estimated that geothermal energy now displaces the equivalent of 4.5 million barrels of oil annually in the United States.

At high temperatures thermal energy is used to produce electricity directly. A major advantage to thermal energy is its reliability. It is consistent for as much as 95 to 100 percent of the time. However, geothermal basins must be managed or they can be depleted in a few decades. Geothermal energy is one of the few renewable energy sources that is a completely reliable source of electricity at any given time.

The major geothermal zones are associated with recent volcanic activities, and major power developments have occurred in Mexico, Central America, New Zealand, Japan, Indonesia, the Philippines, and Italy as well as the United States. There are four principal geothermal resources: hydrothermal, geopressured, hot-dry rock, and magma.

Hydrothermal reservoirs contain hot water or steam trapped in fractured or porous rock. These reserves of energy are enormous. The U.S. Department of Energy estimates the U.S. hydrothermal reservoirs could produce 2,400 quads of energy annually on a sustained basis.

Geopressured reserves are limited. These hot water aquifers contain dissolved methane gas trapped under high pressure in sedimentary rock. The major resource area is the northern Gulf of Mexico along the Texas and Louisiana coasts.

Hot-dry energy resources are accessible geologic formations that are hot but contain little or no water. The highest quality of hot-dry steam is found in Italy and the geysers in northern California. In the geysers, located north of San Francisco, 28 plants generate more than 2,000 megawatts of power at about 4 cents per kilowatt-hour.

Magma is very deep molten rock, a potentially huge source of energy, but little research and development has occurred in this area.

The development of geothermal energy has been modest in recent years. U.S. geothermal power capacity has grown from 78 megawatts in 1968 to more than 800 megawatts in 1980 and to about 2,700 megawatts in 1990.

Technology has been developed to utilize geothermal energy. The reserves are large. It is economical today. Geothermal energy could expand significantly in the immediate future.

Wind

Wind energy was one of the first inanimate forms of energy utilized by man as civilization evolved. In the United States the use of wind power reached its peak in the late nineteenth and early twentieth centuries when small windmills provided farms in the Midwest with the initial electric power for such activities as pumping water and also for the initial use of electricity in homes. As rural electrification developed, most of these small windmills disappeared.

When the energy crisis developed in the 1970s, the industry had progressed little in more than a half century. Basic information was lacking. Data was not available on the resource base, or how the intermittent electric output could be integrated into existing electrical systems.

The development of wind energy has proceeded along two paths. The first was to identify sites where average wind speeds would justify development. A national assessment identified average wind speed greater than 12.5 miles per hour at 33 feet elevation to be found along much of the Pacific, North Atlantic, Great Lakes, and Texas coasts, as well as in large expanses of the Great Plains and in mountain passes and ridges. Most states have a number of desirable sites.

Technical developments have also occurred. The federal R&D program has progressively developed larger machines with the goal of designing large multimegawatt, technically sophisticated turbines for commercial operations. This emphasis on larger machines is due to their greater efficiency, because the power of the wind increases at the cube of its velocity. This means that when the wind speed increases from 10 to 13 miles per hour the available power doubles.

In contrast to the government sponsored programs, private developers have emphasized smaller, simpler turbines. These windmills are primarily for the single-family market, ensuring fully independent electric systems. With the development of the smaller machines, the concept of "wind farms" has developed in which a number of small systems, possibly 20 or more windmills rated at 50 kilowatts each, are clustered on one site, and are treated as a 1-megawatt generating unit.

Utilization of wind power accelerated in the early 1980s. California was the leader in the growth—primarily Altamont, Tehachapi, and San Gargonio—followed by Hawaii. There are now 17,000 wind turbines in operation, with wind power becoming the nation's fourth largest renewable source of electricity (surpassed only by hydropower, biomass, and geothermal energies). With more than 1,500 megawatts of wind power capacity, California and Hawaii produce about 2 billion kilowatt-hours of wind power electricity. It is estimated that the investment totals $2.5 billion and reduces the atmosphere by 11 million pounds of air pollutants and 304,000 tons of carbon each year.

Wind energy costs have decreased from $3,100 per kilowatt in 1981 to an average of $1,000 in 1990, yielding a cost per kilowatt-hour of 7–9 cents, including 1 cent for operations and maintenance. This figure is close to that of fossil fuel operated plants.

The peak of federally sponsored research on windmill technology occurred in the early 1980s. Since then the federal budget has declined from $77 million in 1981 to $9 million in 1989. Federal legislation focused on aerodynamics and structural dynamics. The modern windmills have proven reliable with blade spans ranging from 50 to 100 feet, and power ratings between 100 and 250 kilowatts. Research is increasingly carried out in cooperation with industry. For example, the Solar Energy Research Institute, in conjunction with major windmill operators, is conducting field tests of the world's first computer optimized wind turbine blades.

Wind is still not a significant source of energy. Wind power potential remains undeveloped. A primary reason for lack of development is due to the inconsistency of the wind and the need to store the electricity until there is a demand. It is estimated that if the total wind power sites of California were harnessed they could produce 13,000 megawatts of electricity. This would be about one-fifth of the electricity demand in the state.

Wind power is an energy source with enormous potential. As the nation's largest utility, Pacific Gas and Electric, stated in 1989, "Wind energy technology has improved greatly in the last few years and could be a major source of economically competitive energy without any technical breakthroughs."

Biomass Energy

Biomass has been used for thousands of years to provide energy for ordinary home activities. There are a wide variety of biomass energy sources including wood, plant materials, and animal wastes. Although precise figures are lacking, the U.S. Department of Energy estimated that wood contributed 2.6 quads of energy in 1983, about 3.5 percent of that year's energy use.

Wood is used to a greater extent in rural areas than in urban areas because it is bulky and is relatively low in Btu's. Dry wood has about 17.2 million Btu's per ton, which is less than half that of a ton of fuel oil. Because wood reserves are widely distributed and can be exploited with modest equipment expenditure, and have generally escaped environmental problems, wood consumption rose rapidly after the 1973 energy crisis. Forest products indirectly account for over 60 percent of wood energy used. Overall, approximately 70 percent of the forest product industry's wood fuel occurs in the South, the nation's leading wood-producing region. The South is followed by the West with 17 percent, Northeast with 8 percent, and Midwest with 5 percent.

It is estimated that the residential sector uses about 0.9 quads of wood annually. The U.S. Department of Energy estimates that wood has become the primary heating source in about 6.6 percent of the 83.8 million homes in the United States, and over 17 percent of those homes are in nonmetropolitan counties. Three-quarters of the residential wood fuel is procured directly by the homeowner from nearby forests, generally at no or modest costs.

In recent years the consumption of wood for energy production has extended beyond the home. Several utilities utilize wood to generate electricity. For example, one plant in Vermont burns 500,000 tons of green wood chips per year, producing 50 megawatts of electricity. Central Michigan University has installed a 120-ton-per-day wood chip plant in order to reduce its annual electric bill from $2.5 to less than $1 million. The pulp and paper industry is the largest consumer of wood for energy, meeting 55 percent of its own energy needs with wood and wood processing wastes. The lumber and wood products industry has a similar use of wood.

A growing use of biomass is to supply fuel for generating electricity. More than 320 power plants using biomass as a primary fuel are connected to utility grids. Some of these were built by utilities. Many others generate electricity that is purchased by utilities. In total, there are more than 5,000 megawatts of biomass-fueled generating capacity in the United States.

The use of energy crops such as grasses, shrubs, and grains is the second biomass contribution to the U.S. renewable energy resources in the production of ethyl alcohol. Ethanol can be produced from any vegetable matter. The ethanol product contains approximately 85,000 Btu's per gallon, 38 percent less than gasoline. Extensive plans were initiated during the 1970s and 1980s energy crisis to substitute ethanol for gasoline. The United States has sufficient land to develop the program.

There are a total of 564 million acres of cropland in the United States, of which about 328 million acres are used for growing crops. Forests, pastures, and range cover an additional 158 million acres. In 1988, 78 million acres were taken out of production as part of a crop acreage reduction program. Planting those 78 million acres in energy crops such as corn, which produce 5 tons per acre of feedstock, could yield more than 6 quads of energy annually. Actual production has grown from 8 million gallons in 1978 to 430 million gallons in the mid-1980s. This was approximately 0.5 percent of all gasoline consumed. The development of the ethanol industry has grown slowly because of the difficulty to compete economically. The transportation costs of biomass are high, so that most biomass must be used close to its source. Technological progress in developing energy crops and increasing the efficiency of energy-conversion processes will bring biomass energy costs down. Most ethanol consumption is in the Midwest. While a mixture of gasoline and ethanol (gasohol) accounts for about 5 percent of the gasoline sold nationally, in Iowa it accounted for 32 percent; Nebraska, 26 percent; Kansas, 20 percent; and Indiana, 20 percent. Ethanol has a potential that is still not reached.

Municipal waste landfills provide a major source of potential energy. The United States has the largest number of municipal waste incinerators in the world. The primary function of most of these incinerators is to reduce waste volume, but the waste can become a source of fuel to produce electricity. Most of the waste facilities designed to produce electricity remain in the planning stage.

Garbage also provides a source of methane gas. The use of anaerobic digestion to produce methane from the approximately 18,000 tons of sludge produced by municipal wastewater treatment plants has considerable potential. The rate of growth will depend to a considerable degree on removing constraints from institutional resistance within sewage companies and local governments. The American Gas Association estimates that 0.2 quads of landfill gas can be recovered economically. At the present time nearly 100 facilities recover methane from landfills.

Ocean Energy

The oceans are a potential source of energy that is essentially untapped. Tidal energy uses the force of the incoming-outgoing flow of water in the same manner as hydroelectric plants use the flow of rivers. Each tidal site is specific and the engineering must reflect a design that ensures adequate water flow.

Wave energy technology is still in its infancy. In most systems, waves are forced through a channel to increase pressure. New designs also use floating rafts, buoys, or bobbing floats.

Although there are only a few tidal plants in operation, several exploratory facilities have been built. At La Rance, France, a 240-megawatt facility currently is in operation. Canada has built a plant on the large tidal flats located in the Bay of Fundy. In 1986 the Chinese government erected several small tidal plants, and completed a large 10-megawatt plant in the Zhejiang Province.

Wave energy is more widely applicable for production of small quantities of electricity. Twelve wave energy conversion plants are in operation—11 for electric power generation and one for freshwater production. Cost studies indicate that wave energy can be economical.

Research has also begun in the United States and Japan on Ocean Thermal Energy Conservation (OTEC). This technique is based on the large differences in temperature between ocean surface waters and those at 3,000 feet deep or more. The concept is to pump cold water to the surface to create a solar thermal cycle, or "heat engine." While engineers have investigated the concept of ocean thermal power for the past 50 years, no commercial plants are in operation. A 15-megawatt U.S. prototype exists in Hawaii, and Japan is building a larger plant. Intensive research is needed to improve OTEC plant components and materials and reduce costs, as well as resolve environmental problems. An OTEC system could damage the local ecosystems.

Ocean energy will not be of any importance in supplying energy in the immediate future. Nevertheless ocean energy technology deserves support as a future potential source of energy.

Restraints in Development

A basic question is, why has utilization of renewable energy resources been so slow to develop? A wide variety of barriers have hindered growth. The inherent constraints include:

1. High initial costs for equipment
2. Storage requirements and costs due to the intermittent supply of many renewable sources such as wind and solar
3. Great regional variations in renewable energy
4. Private and public investment decisions are biased because of high initial cost
5. The long-term gamble regarding availability of fossil fuels
6. Reliability of renewable energy equipment
7. Future innovations

There are also federal policy restraints such as:

1. Tax policies
2. Exclusion of social costs because competing fuels exclude such social costs as air, water, land, pollution impacts, national security risks, climate changes, and health costs
3. Subsidies for the development of competing fuels, particularly nuclear energy
4. Marketing and other promotional endeavors, such as rural electrification programs
5. Limit on utility liability and similar programs that subsidize centralized technologies

The market restraints include:

1. The co-evaluation of renewable energy in established energy industries with decades of technology research and market infrastructure development

2. The profit margins are too low to fund large-scale research and development
3. The low energy density of renewable energy in market areas requiring a high energy density

These uses are best met by traditional energy sources. In addition, there are a number of institutional restraints such as:

1. Resistance to change of established practices by the building industry
2. Lack of information reduces market appeal and prevents full assessment of renewable energy potential
3. The legal uncertainties that ensure access to renewable energy, such as ensuring access to sunlight for solar energy

Renewable energy will not provide a significant share of the energy needs in the United States in the immediate future. The use of solar energy for home heating and cooling shows some promise and could be well established by the year 2020. The commercial use of renewable energy may grow by supplying electricity during peak demands. It is possible that within the next century renewable energy will assume a greater role in supplying the nation's energy needs, but even then it will be only one of a number of energy sources that are part of a world system of energy supply and demand.

A National Energy Policy

When the United States experienced its first energy crisis in the early 1970s it was thought that a national energy policy was essential for a sound economic future. Traditionally, energy was in such abundance that development of the energy resources was the responsibility of private industry. Market factors dominated in controlling the output of a particular type of energy.

Although, from time to time, there has been active discussion, neither the federal government nor energy companies have given a long-range national energy policy high priority. The plans introduced by Presidents Nixon, Ford, and Carter reveal that it is easier to prepare broad theoretical alternatives than it is to devise a long-range system based on a complex energy mix. Because the economics of specific energy segments differ so greatly, general principles are difficult to formulate.

The development of a sound energy policy is basically an economic and political process, rather than a technical one. An established energy policy normally visualizes a planned market system for all the energy components. A national energy policy must not be based on the events of a particular moment, but must be based on long-term needs for less than 25 years. Further, to be successful, a policy must be able to adjust to changing situations. This suggests freedom in a changing economy. This also implies an efficiency in the allocation of reserves, and the absence of day-to-day government intrusions into private choices concerning resource allocations.

There is a fundamental need for the United States to develop a sound national energy policy. The energy crisis of the early 1970s demonstrated the devastating problems that can evolve when no policy exists. A handful of developing countries used resources to change U.S. foreign policy and raise oil prices to control the economy. The U.S. government appeared helpless in combating these trends. The country whose development had been shaped by the availability of a rich mineral endowment had to come to terms with rising demand and decreasing resources.

The energy crisis developed temporary interest in establishing a national policy. However, with the disappearance of the crisis, intent waned and there has been little progress in securing self-sufficiency in our energy needs. Since then there has been a detailed administrative history of the crisis year, an analysis of relations between interest groups and government agencies, and many individual energy studies, but there has been little analysis and interpretation in the broadest terms of the way in which our political economy as a whole responds to changing energy patterns.

The efforts of the U.S. government, perhaps well intentioned, have done little to resolve the energy problems raised by changing energy demands. The federal policy process has been dominated by the demands of energy companies seeking to avoid economic losses and, therefore, does not define clearly or promote other objectives, among which are equity, national security, and efficiency. Attempts to regulate the energy market has proved to be counterproductive. Evidence from the energy sector suggests efficiency. Attempts to regulate the energy market has proved to be counterproductive. Evidence from the energy sector suggests that such regulations are increasingly prone to degenerate into a process of adversarial conflict that is damaging to the economic health of the nation. Unfortunately, the present conventional approaches do not address the problems, leaving little opportunity to identify viable alternatives.

In the years before the energy crisis, energy affairs were treated, for the most part, by specific sectors—that is coal, petroleum, natural gas, electricity, and nuclear power. Because of the overlapping of individual energy sectors, they must be treated as a total energy regime. This approach is broader than is usual in policy studies, encompassing private as well as public regulatory systems. Traditionally the energy constituents came to terms on how to govern their share of the energy world. A central concern of all energy companies was the problem of controlling market competition. For each sector this resulted in a different industrial structure and a different pattern of relationships with governmental authorities. Though markets for fuels were, to a significant degree, interdependent, each fuel type tended to develop its own political subsystem in which industry representatives dealt with specialized government counterparts to shape public policies. At no time was the federal government cognizant of, or administratively involved with, the energy system as a whole. There was no coherent attempt to manage or influence the energy regimes as a group. Unfortunately each of the sectors was quite vulnerable to sudden market adjustments, and the capacity of the nation to respond to the crises was heavily influenced by the patterns of sector governance that prevailed.

In order to develop a permanent energy policy, there must be the cooperation of the total energy industry and the government. The benefits from such a policy would provide stability rather than the "boom and bust" that has characterized the energy economy. Even more important, a national policy will provide a means for ensuring that the energy needs of the nation would be met. Energy is the keystone to the U.S. economy. In spite of a recognized need, the task has proved too great to date and a national energy policy remains a myth.

2

Chronologies and Statistics

General Chronologies

Energy

1850 Wood provided 85 percent of the energy consumed in the United States.

1899 Coal provided 71 percent of the nation's energy, wood 20 percent, and natural gas and petroleum about 9 percent.

1929 Coal provided 63 percent of the nation's energy, petroleum and natural gas about 30 percent, and water power and wood the remainder.

1960 Petroleum supplied 36 percent of the nation's energy, natural gas 31 percent, coal 20 percent.

1990 Proved energy reserves of the United States consisted of coal 73 percent, uranium oxide 21.7 percent, shale oil 2.9 percent, natural gas 1.4 percent, and petroleum 1 percent.

1992 Coal supplied 32 percent of the nation's energy, petroleum 25 percent, natural gas 27 percent, uranium 7 percent, and water power 3 percent.

Coal

250–400 M.Y.B.P. Coal first deposited in the Appalachian Fields about 400 million years ago during the Silurian Epoch of the Paleozoic Period.

250–400 M.Y.B.P. *cont.*	Most of the coal deposited about 250 million years ago in the Mississippian and Pennsylvanian Epochs of the Paleozoic Period.
1000 B.C.	Chinese burned coal more than 4,000 years ago.
A.D. 1200	First documented proof of coal mined in Europe occurs in the chronicle of the monk Reinier, of the Priory of St. Jacques at Liege, Belgium.
13th Century	Mining operations developed in England, Scotland, and in a number of fields in the European continent.
13th–14th Centuries	Early prejudice against use of coal in Europe due to noxious fumes. During the reign of Edward I (1239–1307) the death penalty was imposed on any one found guilty of burning coal.
15th–16th Centuries	Increased use of coal in brick kilns led to a wider demand for coal.
18th–19th Centuries	Coal used in iron smelting in Great Britain and Europe provided the foundation for widespread expansion in the metallurgical and engineering industries.
1701	Earliest record of coal mined in the United States on James River, near Richmond, Virginia.
1755	Coal mined in Ohio by early settlers.
1758	First recorded commercial shipment of 32 tons of coal from the Appalachian fields to Norfolk, Virginia, and then by boat to New York.
1800	Coal production in the United States exceeds 108,000 tons.
1835	Bituminous coal production in the United States exceeded 1,000,000 tons.
1852	First commercial shipment from the famous Pittsburgh Coal Seam near Fairmont, West Virginia.
1840–1860	More than three-fourths of the bituminous coal production in the United States came from Pennsylvania.
1880–1918	Anthracite production rose at a slower rate increasing from 31.9 millions tons in 1880 to 98.8 million tons, an all time peak.

1890–1918 Bituminous coal emerged as an important residential and industrial fuel with output rising from 212 million tons in 1900, to 417 million tons in 1910. In 1918, bituminous coal reached its initial peak with an output of 579 tons.

1920s Petroleum and natural gas become highly competitive with coal.

1937 In order to improve the economic conditions of the coal industry, Congress passed the Bituminous Coal Act (Guffey Act). A price regulation system was enacted. It remained in effect until the coal industry recovered during World War II. The act was abolished in 1943.

1939–1947 Bituminous coal production increased from 394 million tons to 619 million tons due to shortages of petroleum and demand for coal in the war efforts.

1947–1961 Bituminous coal production declined to a modern low of 402 million tons.

1961–1970 With increasing demand by the electric utility industry, bituminous coal recovered in the 1960s with production reaching 602 million tons in 1970.

1970 Clean Air Act enacted by Congress. High sulfur coals could not be burned until sulfur content reduced. The coal industry east of the Mississippi River was depressed because of the high sulfur content. Yet, coal production soon increased in western United States due to the availability of low sulfur coals.

1973 Oil embargo revealed the need to help develop domestic energy sources. Coal industries with the highest reserves increased production.

1990 Bituminous coal production exceeded one billion tons for first time. Wyoming was the leading coal producing state.

Petroleum

2000 B.C. Petroleum has been used for medicinal purposes since the onward beginning of civilization. Surface oil seeps are found throughout the world.

1859 The first well was drilled to produce petroleum at Titusville, Pennsylvania. Oil production began on August 27.

1900s By 1900 many wells had ceased to produce or were onward producing at low levels although only a small fraction of the oil had been produced from the oil horizon. To recover the remaining oil, secondary techniques were developed. This usually consisted of pumping water, gas or air into the producing horizon. This process reestablished the original reservoir pressure causing oil to once again flow into the producing well.

1910–1955 Annual production increased from 209 million barrels to 2,500 million barrels.

1922 Tetraethyl lead first used in gasoline to reduce engine knock and increase power.

1950s Demand for oil was increasing rapidly and American companies began to seek the major oil deposits throughout the world, particularly in the Middle East and Venezuela.

1960 Organization for Petroleum Exporting Countries (OPEC) created by major exporting countries to gain control of their domestic industry.

1967–1973 Petroleum imports to the United States increased from 2.2 million barrels per day to 6 million barrels per day.

1973 OPEC nations imposed an embargo of oil on the United States.

1973–1980 Price of a barrel of oil rose from $2.23 to over $41.00.

1970s U.S. government attempted to develop a policy to decrease the reliance of the nation on foreign oil. These policies were temporary measures and no national energy policy has evolved.

1980s Energy supplies have been adequate in the 1980s and the Reagan administration promoted conventional energy production, attempting to remove regulations and legal obstacles, and enlarged access to energy sources.

Uranium

1789 The element uranium was discovered by Martin Heinrich Klaproth in a pitchblende mineral mined in Saxony, Germany.

1938 Otto Hahn and Fritz Strassman demonstrated that bombardment with neutrons caused uranium to break down into radioactive isotopes releasing energy.

1940s   Major mining operations began for uraninite, uranium dioxide, in the plateau areas of Colorado, Utah, New Mexico, and Arizona.

1945   First atomic bomb was detonated in Nevada on July 16, 1945.

1948   August 5, Atomic Energy Commission (AEC) establishes a Division of Reactor Development.

1951   December 20, first generation of electric power from a nuclear reactor at AEC's reactor testing center in Arco, Idaho.

1953   December 8, President Dwight Eisenhower delivers "Atoms for Peace" speech before the United States.

1954   March 14, AEC and Duquesne Light Company negotiate with an agreement to construct jointly a pressurized water reactor demonstration facility at Shippingport, Pennsylvania.

1955   March 30, AEC created a Division of Licensing for nuclear energy.

1956   January 6, Power Reactor Development Company (PRDC) applies to AEC for a construction permit to build a commercial fast-breed reactor in Lagoona Beach, Michigan.

1957   The world's first generating nuclear power plant began operating at Shippingport, Pennsylvania.

1973   French government announces plans to produce 70 percent of the nation's electricity and 30 percent of the total consumption of energy of the nation by nuclear power by 1990.

1979   March 28, Three Mile Island accident at Middletown, near Harrisburg, Pennsylvania.

1987   April 25, Chernobyl nuclear explosion.

1989   The world's first nuclear power plant at Shippingport, Pennsylvania, is dismantled.

Chronologies of Specific Energy Sources

In the twentieth century the energy system of the nation consists of a number of components. The following critical dates reveal not only the different types of energy, but also the varying importance of each. For ease of use, seven chronologies are presented: Energy, Petroleum, Natural Gas, Coal, Nuclear Energy, Renewable Energy, Laws and Regulations.

Energy

Table 2.1
World Primary Energy Production (Quadrillion Btu)

	1975	1980	1985	1990
North America	71.06	80.48	84.09	87.98
United States	59.73	64.64	64.58	67.47
Central/South America	10.62	12.15	13.42	16.46
Venezuela	5.90	5.71	4.77	6.29
Western Europe	21.36	28.65	34.98	36.54
West Germany	4.95	5.44	6.10	5.77
United Kingdom	4.47	8.40	10.11	8.66
Middle East	43.58	42.13	25.66	41.04
Saudi Arabia	15.68	22.48	8.55	15.81
Iran	12.23	3.91	5.57	7.65
Africa	13.19	17.95	18.42	21.45
Nigeria	3.88	4.50	3.35	4.14
Far East-Oceania	29.71	36.03	48.99	60.17
China	15.11	18.33	25.14	30.23
Eastern Europe/U.S.S.R	56.51	68.20	83.54	81.43
U.S.S.R	43.62	55.67	68.49	67.64

Source: *Annual Energy Review.* Washington, D.C.: Energy Information Administration, Office of Energy Markets and End Use, U. S. Department of Energy.

Table 2.2
United States Energy Production and Consumption
(Trillion Btu)

	Production	Consumption
1850	2,354	2,357
1870	3,967	3,952
1890	7,134	6,990
1910	16,601	16,026
1930	22,822	22,961
1950	34,101	33,656
1970	62,100	66,400
1980	64,800	76,000
1990	67,600	81,500

Source: *Annual Energy Review.* Washington, D.C.: Energy Information Administration, Office of Energy Markets and End Use, U.S. Department of Energy.

Table 2.3
United States Energy Production by Type
(Trillion Btu)

	Coal	Petroleum	Natural Gas	Hydro-power	Nuclear Power	Total Domestic Gross Energy Imports
1950	14,647	11,449	6,841	1,415	—	34,352
1960	11,140	14,664	14,135	1,608	6	41,553
1970	15,067	20,401	24,154	2,630	229	62,481
1980	18,597	18,249	22,162	2,900	2,730	64,761
1991	21,552	15,609	20,713	2,880	6,543	67,488

Source: *Annual Energy Review.* Washington, D.C.: Energy Information Administration, Office of Energy Markets and End Use, U.S. Department of Energy.

Table 2.4
United States Energy Consumption by Sectors
(Trillion Btu)

	Household and Commercial	Industrial	Transporation	Electricity	Misc.	Total Consumption
1950	7,593	12,325	8,616	4,981	477	33,992
1960	10,174	14,642	10,818	8,263	672	Mw4,569
1970	13,988	20,225	16,473	16,242	215	67,143
1980	10,720	21,072	19,658	24,505	NA	75,955
1990	9,939	19,921	22,009	29,578	NA	81,453

Source: *Annual Energy Review.* Washington, D.C.: Energy Information Administration, Office of Energy Markets and End Use, U.S. Department of Energy.

Table 2.5
Selected United States Economic, Demographic, and Energy Indicators

	Gross Domestic Product (Bill. $ 1987)	Population (Million)	Gross Energy Input (Quadrillion Btu's)	Gross Energy Consumption Per $ of 1987 GNP (1,000 Btu's)	Gross Energy Consumption Per Capita (Mill. Btu's)
1950	NA	152.3	33.1	NA	217.3
1960	1,973	180.7	43.8	22.2	242.4
1970	2,876	205.1	66.4	23.1	323.7
1980	3,776	227.7	76.0	20.1	333.8
1991	4,849	252.2	81.5	16.8	323.2

Source: *Basic Petroleum Data Book.* Annual. Washington, D.C.: American Petroleum Institute.

Petroleum

Table 2.6
Estimated Proved World Crude Oil Reserves
(Annually as of January 1; Millions of Barrels)

	Africa	Austral-asia	Canada	Latin America	Middle East	Western Europe	Soviet Union	United States	Total World
1950	202.5	1,557.0	1,200.0	11,400.8	32,413.0	289.0	4,741.0	24,649.5	76,452.7
1955	112.0	2,708.0	2,207.6	14,356.0	97,459.0	1,049.2	10,047.0	29,560.7	159,499.6
1960	7,273.5	10,175.5	3,497.1	24,435.6	131,436.0	1,496.0	30,002.0	31,719.3	290,035.1
1965	19,395.0	11,613.7	6,177.5	25,525.0	212,180.0	2,035.6	30,750.0	30,990.5	338,668.0
1970	54,679.5	13,138.2	8,619.8	29,179.8	333,506.0	1,779.0	60,000.0	29,631.9	530,534.1
1975	68,299.5	21,047.7	7,171.2	40,578.0	403,858.2	25,814.0	111,400.0	34,250.0	712,418.5
1980	57,072.1	19,355.2	6,800.0	56,472.5	361,947.3	23,476.4	90,000.0	29,810.0	644,933.0
1985	55,540.6	18,529.9	7,075.0	83,315.7	398,380.8	24,425.5	84,100.0	28,466.0	699,813.4
1990	58,836.6	22,545.4	6,133.5	125,026.9	660,247.2	18,822.1	84,100.0	26,501.0	1,002,212.7
1992	60,487.5	44,073.2	5,587.8	119,765.5	661,570.8	14,503.2	NA	26,250.0	991,011.4

Source: *Basic Petroleum Data Book.* Annual. Washington, D.C.: American Petroleum Institute.

Table 2.7
Estimated Proved Crude Oil Reserves
(Thousands of Barrels)

Year	Country	Reserves
1975	World	712,418,535
	Saudi Arabia	164,500,000
	U.S.S.R.	83,400,000
	Kuwait	72,800,000
	Iran	66,000,000
	Iraq	35,000,000
	United States	34,249,906
	United Arab Emirates	33,920,000
	Libya	26,600,000
	China	25,000,000
	Nigeria	20,900,000
1980	World	644,933,000
	Saudi Arabia	163,350,000
	U.S.S.R.	67,000,000
	Kuwait	65,400,000
	Iran	58,000,000
	Mexico	31,250,000
	Iraq	31,000,000
	United States	29,810,000
	United Arab Emirates	29,411,300
	Libya	23,500,000
	China	20,000,000
1985	World	699,813,400
	Saudi Arabia	169,000,000
	Kuwait	90,000,000
	U.S.S.R.	63,000,000
	Mexico	48,600,000
	Iran	48,500,000
	Iraq	44,500,000
	United Arab Emirates	32,390,000
	United States	28,446,000
	Venezuela	25,845,000
	Libya	21,100,000
1992	World	991,011,400
	Saudi Arabia	257,842,000
	Iraq	100,000,000
	United Arab Emirates	97,700,000
	Kuwait	94,000,000
	Iran	92,860,000
	Venezuela	59,100,000
	U.S.S.R.	57,000,000
	Mexico	51,298,000
	United States	26,250,000
	China	24,000,000

Source: *Basic Petroleum Data Book.* Annual. Washington, D.C.: American Petroleum Institute.

Major Petroleum Provinces: Discovery Dates

1859 Appalachian Province. First oil well drilled at Titusville, Pennsylvania. Oil discovered on August 27, 1859

1884 Lima–Indiana Province

1887 California Province

1900 Gulf Coast Province

1905 Illinois Province

Midcontinent Province

1925 Michigan Province

1951 Northern Great Plains Province

1967 Northern Alaska

U.S. Crude Oil Production and Imports

Table 2.8
Crude Petroleum Production by States
(42-gallon barrels)

1950	World	3,796,658,000
	U.S.	1,971,845,000
	Texas	829,231,000
	California	327,627,000
	Louisiana	209,116,000
	Oklahoma	164,899,000
	Kansas	107,506,000
	Illinois	61,922,000
	Wyoming	60,457,000

1960	World	7,684,752,000
	U.S.	2,574,933,000
	Texas	933,632,000
	Louisiana	394,360,000
	California	304,356,000
	Oklahoma	192,288,000
	Wyoming	135,521,000
	Kansas	113,455,000
	New Mexico	107,940,000
1970	World	16,689,617,000
	U.S.	3,517,450,000
	Texas	1,249,697,000
	Louisiana	906,907,000
	California	372,191,000
	Oklahoma	223,574,000
	Wyoming	160,345,000
	New Mexico	128,184,000
	Kansas	84,853,000
	Alaska	83,616,000
1980	World	21,760,880,000
	U.S.	3,146,400,000
	Texas	977,436,000
	Alaska	591,646,000
	Louisiana	469,141,000
	California	356,923,000
	Oklahoma	150,140,000
	Wyoming	126,362,000
	New Mexico	75,324,000
	Kansas	60,151,000
1991	World	21,464,024,000
	U.S.	2,707,000,000
	Texas	906,961,000
	Alaska	656,349,000
	Louisiana	414,469,000
	California	351,016,000
	Oklahoma	108,095,000
	Wyoming	99,928,000
	New Mexico	70,416,000

Sources: *Basic Petroleum Data Book.* Annual. Washington, D.C.: American Petroleum Institute; *Minerals Yearbook.* Annual. Washington, D.C.: Bureau of Mines, U.S. Department of Interior.

Table 2.9
Total U.S. Petroleum Imports from OPEC Nations
(Thousands of Barrels per Day)

1960 - 1,815		1976 - 7,313
1961 - 1,917		1977 - 8,807
1962 - 2,082		1978 - 8,363
1963 - 2,123		1979 - 8,456
1964 - 2,258		1980 - 6,909
1965 - 2,468		1981 - 5,996
1966 - 2,577		1982 - 5,113
1967 - 2,537		1983 - 5,051
1968 - 2,840		1984 - 5,437
1969 - 3,168		1985 - 5,067
1970 - 3,419		1986 - 6,224
1971 - 3,925		1987 - 6,678
1972 - 4,741		1988 - 7,402
1973 - 6,251		1989 - 8,061
1974 - 6,112		1990 - 8,018
1975 - 6,056		1991 - 7,627

Source: *Basic Petroleum Data Book.* Annual. Washington, D.C.: American Petroleum Institute.

Table 2.10
Share of Imports in United States Domestic
Production Demand
(Thousands of Barrels)

	Imports	Domestic Demand for Refined Products	Import as a % of Demand
1950	310,261	2,392,974	13.0
1960	664,111	3,585,820	18.5
1970	1,248,062	5,364,473	23.3
1980	2,512,636	6,242,445	40.3
1991	2,783,763	6,100,550	45.6

Source: *Basic Petroleum Data Book.* Annual. Washington, D.C.: American Petroleum Institute.

Table 2.11
United States Imports of Crude Oil
(Thousands of Barrels)

1950	U.S.	177,714	
Western Hemisphere			136,096
	Venezuela	107,019	
	Colombia	16,159	
Middle East			41,618
	Kuwait	26,741	
	Saudi Arabia	14,650	

1960	U.S.	371,575	
Western Hemisphere			230,229
	Venezuela	172,887	
	Canada	41,349	
	Colombia	14,799	
Middle East			113,175
	Kuwait	47,512	
	Saudi Arabia	28,232	
	Iran	13,056	
Far East			26,720
	Indonesia	26,720	
Africa			1,451

1970	U.S.	483,293	
Western Hemisphere			351,366
	Canada	245,258	
	Venezuela	97,996	
Middle East			61,892
	United Arab Emirates	23,047	
	Iran	12,184	
	Kuwait	12,123	
Africa			44,365
	Nigeria	17,490	
	Libya	17,156	
Far East			25,670
	Indonesia	25,670	

1980	U.S.	1,926,162	
Africa			731,342
	Nigeria	307,840	
	Libya	200,406	
	Algeria	166,980	
Middle East			561,248
	Saudi Arabia	457,671	
	United Arab Emirates	63,052	
	Iraq	10,328	
Western Hemisphere			379,813
	Mexico	185,541	
	Canada	73,002	
	Venezuela	56,950	
Far East			137,551
	Indonesia	137,551	
Europe			116,208
	United Kingdom	63,459	
1991	U.S.	2,110,532	
Western Hemisphere			806,317
	Mexico	276,864	
	Canada	271,375	
	Venezuela	243,986	
Middle East			646,204
	Saudi Arabia	621,766	
	Iran	11,821	
Africa			423,148
	Nigeria	249,403	
	Angola	92,839	
	Gabon	30,556	
Far East			87,412
	Indonesia	37,231	
Europe			66,945
	United Kingdom	38,862	
	Norway	27,009	

Source: *Basic Petroleum Data Book.* Annual. Washington, D.C.: American Petroleum Institute.

Table 2.12
Average Annual Wellhead Price of United States Crude Oil
(U.S. Dollars per U.S. Barrels of 42 Gallons)

	Price in Current Dollars	Price in 1982 Constant Dollars
1960	2.88	9.09
1961	2.89	9.15
1962	2.90	9.15
1963	2.89	9.15
1964	2.88	9.11
1965	2.86	8.85
1966	2.88	8.65
1967	2.92	8.74
1968	2.94	8.60
1969	3.09	8.68
1970	3.18	8.62
1971	3.39	8.90
1972	3.39	8.52
1973	3.89	8.64
1974	6.74	12.60
1975	7.56	12.95

1976	8.14	13.32
1977	8.57	13.20
1978	8.96	12.82
1979	12.51	15.90
1980	21.59	24.04
1981	31.77	32.42
1982	28.52	28.52
1983	26.19	25.85
1984	25.88	24.96
1985	24.09	23.34
1986	12.51	12.49
1987	15.40	14.98
1988	12.58	11.77
1989	15.86	14.14
1990	20.03	17.22
1991	16.50	14.16

Sources: *Statistical Abstract of the United States.* Annual. Washington, D.C.: Bureau of the Census, U.S. Department of Commerce; *Basic Petroleum Data Book.* Annual. Washington, D.C.: American Petroleum Institute.

Major Petroleum Companies

Standard Oil Company and Trust

1863 John D. Rockefeller joined Maurice B. Clark and Samuel Andrews in a Cleveland refining business.

1865 Rockefeller bought out Clark.

1870 Rockefeller brought in Henry M. Flagler.

Rockefeller, Anderson, and Flagler operated largest refineries in Cleveland.

Standard Oil Company incorporated in Ohio.

1880   Controlled refining of 90 to 95 percent of all oil produced in United States.

1882   Standard Oil Company and affiliated companies created Standard Oil Trust.

1892   Ohio Supreme Court ordered the trust dissolved but it still operated in New York City.

1899   Standard Oil Company of New Jersey was incorporated as a holding company.

1906   U.S. government brought suit against Standard Oil Company of New Jersey.

1911   Company was ordered to divest itself of its holdings, however, retained Standard Oil in their names after dissolution.

Mobil Oil Corporation

Location, Chairman and Chief Executive Officer
Fairfax, VA
Allen E. Murray

Economic Structure
Employees   58,000
Revenue $56.37 billion

1866   Vacuum Oil Company was founded.

1882   Became part of Standard Oil Trust.

                Standard Oil Company of New York (Socony was another predecessor).

1911   Both companies became independent.

1931   Both companies merged to form Socony-Vacuum Corporation.

1934   Changed name to Socony-Vacuum Oil Company,Inc.

1955   Changed name to Socony Mobil Oil Company, Inc.

1966   Changed name to Mobil Oil Corporation.

1974   Acquired 54 percent of voting shares of Marcor Inc. (parent company of Container Corporation of America and Montgomery Ward & Co.).

1976   Mobil Corporation was formed with merger of Mobil Oil Corporation and Marcor Inc.

1986   Mobil sold Container Corporation of America.

1988   Mobil sold Montgomery Ward & Co.

Operations

Crude petroleum production; natural gas production; crude petroleum pipelines; refined petroleum pipelines; deep sea foreign transportation of freight; petroleum refining; filling stations, gasoline, chemicals and chemical products.

Subsidiaries

Lake Forrest Oasis Mobil (gasoline service stations)
Mobil Chemical Company Division
Mobil Chemical Film Division
Mobil Oil Company
Mobil Oil Pipe Line
Southeast Exploration and Production Division (DE)
and others

Atlantic Richfield Company (ARCO)

Location, President and Chief Executive Officer

Los Angeles, CA
Robert E. Wycoff

Economic Structure

Employees	26,600
Revenue	$16.80 billion

1870   Incorporated as Atlantic Refining.

1892   Part of Standard Oil Trust in eastern United States.

1911   Became independent after dissolution of the Standard Oil Trust.

Richfield Company founded.

1916   Sinclair Oil Company founded by Harry F. Sinclair who made his first big oil strike in Oklahoma.

1966   Atlantic Refining Company and Richfield Oil Corporation merged.

1969   Sinclair Oil Corporation was brought into Atlantic Richfield Company.

1992   Petroleum operations in all parts of the United States, in Indonesia, the North Sea, and South China Sea.

Operates transportation facilities for liquid petroleum.

Operations

Exploration, development, production, purchase, transportation and sale of chemicals; production and marketing of coal, chemicals, and metal products; manufacturing, transportation and marketing of petroleum products derived from crude oil, including petro chemicals; development of solar energy and energy-related activities.

Subsidiaries

Anaconda Company but merged in 1981 into Atlantic Richfield
ARCO Coal Company
ARCO International Oil and Gas Company
ARCO Products Company
ARCO Alaska Inc.
ARCO Oil and Gas Company
ARCO Transportation Company
ARCO Chemical Company
Elwood Onshore Facility
Lyondell Petrochemical Company

Continental Oil Company (Conoco)

Location, President and Chief Executive Officer

Houston, TX
C. S. Nicandros

Economic Structure

Employees   43,141
Revenue $1.55 billion

1875   Founded as Continental Oil and Transportation Company in Ogden, Utah.

1877   Reincorporated in California.

1885	Reincorporated with new name, Continental Oil Company as part of Standard Oil Trust.
1913	Became independently incorporated.
1913–1929	Became fully integrated oil company.
1929	Conoco had 1,800 producing wells.
	Merged with Marland Oil Company with wells and marketing operations from Oklahoma to Maryland.
1946	Acquired refineries or fields in Louisiana, Canada, Libya, Dubai, the North Sea, and Indonesia.
1966	Acquired Consolidation Coal Company.
1981	Acquired by E.I Du Pont de Nemours & Company and reorganized as the Continental Group.

Operations

Crude oil and natural gas production; refined petroleum products, chemical and petrochemicals; ventured into uranium and copper mining.

Subsidiaries

None

Chevron Corporation

Location, Chairman and Chief Executive Officer

San Francisco, CA
K.T. Derr

Economic Structure

Employees	54,800
Revenue	$42.57 billion

1879	Originated with founding of Pacific Coast Oil Company.
1900	Standard Oil Company purchased Pacific Coast Oil.
1906	Merged with West Coast operations to form Standard Oil Company of California.

1911   Standard Oil Company of California became an autonomous entity when U.S. Supreme Court dissolved the giant New Jersey based Standard Oil combine.

1926   Merged with Pacific Oil Company.

1930s   Discovered oil in Bahrain and Saudi Arabia.

1936   Formed marketing enterprise—the Caltex Group Companies—owned by Standard Oil of California and Texaco.

1939   Began operations in Louisiana and offshore in Gulf of Mexico.

1941   Canadian production began.

1961   Purchased Standard Oil Company of Kentucky.

1984   Merged with Gulf Oil Corporation.

Operations

Gasoline; jet fuel and lubricating oils; asphalt; natural gas; chemicals and fertilizers production.

Subsidiaries

Chevron U.S.A. Inc. (PA)
Chevron Capital U.S.A. Inc. (DE)
Chevron Chemical Company (DE)
Chevron Industries Inc. (DE)
Chevron Investment Management Company (DE)
Chevron Land and Development Company (DE)
Chevron Oil Field Research Company (DE)
Chevron Oil Finance Company (DE)
Chevron Pipe Line Company (DE)
Chevron Research and Technology Company (DE)
Huntington Beach Company (CA)
The Pittsburg & Midway Coal Mining Company (MO)
Also subsidiaries in Canada and foreign areas.

Exxon Corporation

Location, President and Chief Executive Officer

Irving, TX
Lawrence G. Rawl

Economic Structure

Employees   144,000
Revenue $105.52 billion

1882	Founded by Standard Oil Trust.
1888	Organized Anglo-American Oil Co. to market oil in British Isles.
1890	Acquired interest in Esso AG, a German firm.
1898	Gained control of Canada's Imperial Oil Limited.
1899	Became holding company for all companies in the trust.
1911	U.S. Supreme Court ordered Exxon to divest itself of 33 of its American subsidiaries.
1919	Acquired Humble Oil & Refinery Company.
1920	Acquired Tropical Oil Company of Colombia.
1921	Acquired Standard Oil Company of Venezuela.
1926	New Jersey company introduced trade-name Esso but not well accepted in several states.
1928	Acquired Crude Petroleum Company of Venezuela.
	Acquired Turkish (now Iraq) Petroleum Company.
1948	Acquired Arabian American Oil Company.
1972	Standard Oil Company (New Jersey) became Exxon Corporation and was accepted by its subsidiaries and affiliates. Foreign affiliates retained Esso name.

Operations

Petroleum production; natural gas production; natural gas liquids production; petroleum refining; deep sea foreign transportation of freight; petroleum bulk stations and terminals; bituminous coal and lignite.

Subsidiaries

Exxon Coal U.S.A. Inc. (DE)
Exxon Pipeline Company (DE)
Exxon Overseas Corporation (DE)
Exxon Coal U.S.A. (DE)
Exxon Capital Holding Corp. (NJ)
Exxon Shipping Company (DE)
Exxon Production Research Company (DE) and several others. Many foreign subsidiaries still use Esso in their

name such as Esso Holding Company.
U.K., Inc. (DE) and many others.

Sun Company, Inc.

Location, President and Chief Executive Officer
Radnor, PA
Robert H. Campbell

Economic Structure

Employees	21,309
Revenue	$9.80 billion

1886   Formed as Sun Oil Company of Ohio.

1901   New Jersey oil and gas business incorporated.

1971   Company was incorporated.

1975   Sun Company was restructured into 14 separate units to
expand into areas not related to its petroleum business.

Uses trade names of Sunoco and DX in selling products.

Operations
Holding company (energy resources); petroleum refining;
petroleum terminals; gasoline service stations; deep sea
foreign transportation of freight; coastwise transportation
freight; crude petroleum pipelines.

Subsidiaries
Sun International Exploration and Production Co. Ltd. (DE)
Atlantic Petroleum Corp. (Netherlands)
Sun Refining and Marketing Co. (PA)
Sun Coal Company
Radnor Corp. (PA)
Suncor Inc. (Canada)
Helois Capital Corp. (DE)

Marathon Oil Company

Location, President and Chief Executive Officer
Pittsburgh, PA
Victor G. Beghini

Economic Structure

Employees	9,250
Revenue	$13.83 billion

1887   Founded as Ohio Oil Company.

1889   Controlled by Standard Oil Trust.

1911   Independent when U.S. Supreme Court broke up the Standard Oil combine.

1962   Adopted name Marathon Oil Company.

1976   Acquired Pan Ocean Oil Corporation and developed interests in minerals such as coal, fluorspar, and zinc.

1981   Became subsidiary of USX Corp.

Operations

Owns and operates domestic and foreign oil and gas exploratory and producing properties and related facilities and equipment; crude oil refining, purchasing, trading, transportation, distribution and marketing; provides administrative and staff support service for its subsidiaries;owns and operates company research center, petroleum bulk stations and petroleum terminals.

Subsidiaries

None.

Amoco Corporation

Location, President and Chief Executive Officer

Chicago, IL.
H. Laurance Fuller

Economic Structure

Employees	54,524
Revenue	$31.58 billion

1889   American Petroleum Corporation was founded by Standard Oil Trust.

1911   U.S. Supreme Court dissolved the Standard Oil Trust.

Divested from Standard Oil Company of Indiana becoming an independent company.

1925   Acquired interests in oil fields in Mexico and Venezuela.

1931   Was recognized under Indiana General Corporation Act.

1949   Acquired interest in oil fields in Canada.

1954   Acquired entire assets of American Oil Company.

1962   Created American International Oil Company as subsidiary.

1985   Present name was adopted.

Operations

Petroleum refining; petroleum bulk stations and terminals; filling stations; gasoline; crude petroleum pipelines; refined petroleum pipelines; styrene resins; polystyrene resins; polypropylene resins; jet fuels. Retail trademark is Amoco.

Subsidiaries

American International Oil Company
Amoco Company (DE)
Amoco Minerals PNG Company (DE)
Amoco Development Company (DE)
Amoco Pipeline Company (ME)
AmProp Finance Company (IN) and many others

Pennzoil Company

Location, President and Chief Executive Officer

Houston,TX
James L. Pate

Economic Structure

Employees	11,600
Revenue	$21.80 billion

1889   Founded as South Penn Oil Co.

1911   Became independent company when Standard Oil Trust was dissolved.

1924   Four companies merged under the name The Pennzoil Company.

1925   South Penn Oil Co. acquired a controlling interest in Pennzoil and operations began nationwide.

1963   Zapata Petroleum and Stetco Petroleum merged with South Penn Oil to form a new Pennzoil Company.

1965   Acquired controlling interest in United Gas Corp.

1968   Pennzoil was consolidated with United Gas to form Pennzoil United Inc.

1972   Renamed Pennzoil Company.

1984   J. Paul Getty showed interest in merging with Pennzoil but reneged.

1985   Pennzoil lawsuit vs. Getty. Pennzoil was awarded $7.53 billion in actual damages and $3 billion in punitive damages.

1988   Texaco ultimately paid $3 billion in cash as final settlement of the Getty-Pennzoil lawsuit.

Operations

Petroleum refining; petroleum bulk stations; filling stations; gasoline filters; oil, fuel and air motor vehicles; petroleum production; natural gas production; sulphur mining.

Subsidiaries

Duval Sales International, S.A. (Belgium)
Jiffy Lube International, Inc. (NV)
Pennzoil Exploration and Production Co. (DE)
Pennzoil Products Co. (DE)
Richland Development Corp. (NV)

Gulf Oil Corporation

Location, President and Chief Executive Officer
See Chevron Corporation

Economic Structure
See Chevron Corporation

1901   Began with the gusher at Spindletop Hill near Beaumont, Texas.

1907   Gulf Oil was incorporated.

1913   Opened first drive-in filling station in Pittsburgh.

1923   The Port Arthur facility was the largest in the world.

1984   Acquired by Chevron Corporation.

1985   Chevron U.S.A. Inc. was merged into Gulf Oil Corporation and name was changed to Chevron U.S.A. Inc.

Operations

Has heavy investment in forms of energy; secondary interests in chemicals, minerals, and nuclear power.

Subsidiaries

None.

Texaco, Inc.

Location, President and Chief Executive Officer

White Plains, NY
James W. Kinnear

Economic Structure

Employees	54,481
Revenue	$41.82 billion

1902   Founded by Joseph S. Cullinan, a former Standard Oil field worker, and Arnold Schlaet, a New York investment manager.

1910   Operated tankers out of New York.

1911   Established first refinery outside Texas in Illinois.

1928   First company to sell products in all 48 states.

1930s   Began operations in Canada, Colombia, and Venezuela. Joined with Standard Oil of California in half ownership of operations in Indonesia and the Middle East.

1940   Sponsored the Texaco-Metropolitan Opera Radio Network in United States and Canada.

1959   Current name was first used.

1980s   Competed with Pennzoil Company to control Getty Oil Company, with Texaco the winner. After a five-year court battle Pennzoil was awarded $10,530,000,000 in punitive damages but Texaco paid only a part of it.

1990   Texaco-Metropolitan Opera International Radio Network extended to Europe.

Operations

Crude petroleum production; natural gas production; natural gas liquids production; crude petroleum pipeline; refined petroleum pipelines; deep sea foreign transportation of freight; petroleum refining.

Subsidiaries

Four Star Oil and Gas Co.
Getty Oil Co.
Texaco Pipeline Inc.
Texaco Cogeneration Co.
Texaco Exploration and Production Inc.
Texaco Refining Marketing (East) Inc.
Texaco Petroleum Co.
Texaco Refining Marketing Holdings Inc.
Texaco Refining & Marketing Inc.
Texaco Trading and Transportation Co.
Texaco TPC Inc.
Texaco Brasil S.A. Productos de Petroleo (Brazil)
Texaco International Trader Inc.
Texaco A/S (Denmark)
Texaco North Sea U.K. Co., (Eng.)
Texaco Limited (Eng.)
Texaco Britain Ltd. (Eng.)
Texaco Denmark Inc. (Denmark)
Texaco Investments (Netherlands) Inc.
Texaco Panama Inc. (Panama)
Texaco Chemical Co.
Texaco Overseas Holdings Inc.

British Petroleum Company (BP)

Location, Chairman and Chief Executive Officer

London, England
Robert Horton

Economic Structure

Employees	119,600
Revenue	$63.72 billion

1909	Anglo-Persian Oil Company was formed.
1912	First cargo of oil was exported from Abadan.
1914	British government principal stockholder.

1933   Concession was revised.

1935   Renamed the Anglo-Iranian Oil Company, Ltd.

1938   Other Iranian fields and refineries built.

1951–1953   Concession suspended.

1953   Concession renewed.

1954   Changed name to British Petroleum Company, Ltd.

1955   British Petroleum became holding company.

1965   Found commercial natural gas in U.K. sector of North Sea.

1970   First discovery of major oil field.

                Merged with Standard Oil Company (Ohio).

1980s   BP was privately owned.

1982   Named the British Petroleum Company.

1987   Acquired all of Standard Oil Company for $8,000,000,000 making it one of the largest oil companies in the world.

Operations

Oil and gas exploration; production; refining and sales; transportation and distribution of oil and natural gas; manufacturing of chemicals, plastics, synthetic fibres, and animal-feed products.

Subsidiaries

BP Chemicals (International) (England)
BP Exploration (Scotland)
BP International (England)
BP Exploration Operating Company (England)
BP Shipping (England)
BP Capital (England)
BP Chemicals (England)
BP Oil UK (England)
BP Austria (Austria)
BP France (France)
Deutsche BP (Germany)
BP Greece (England)
BP Capital (Netherlands)
BP Nederland (Netherlands)

BP Nutritional International (Netherlands)
BP Portuguesa (Portugal)
BP Espana (Spain)
Svenska BP (Sweden)
BP Switzerland (Switzerland)
BP Petrolleri (Turkey)
BP Middle East (England)
BP Southern Africa (South Africa)
BP Singapore (Singapore)
BP Australia (Australia)
BP Developments Australia (Australia)
BP Oil New Zealand (New Zealand)
BP America Standard Oil (USA)
Standard Oil Company

Phillips Petroleum Company

Location, President and Chief Executive Officer
Bartlesville, OK.
W. W. Allen

Economic Structure

Employees	22,400
Revenue	$13.60 billion

1917   Incorporated to acquire the oil producing properties and businesses in Oklahoma and Kansas of Frank Phillips and L. E. Phillips.

1927   Acquired refinery that led to development of integrated oil company.

Operations
Crude oil; natural gas; liquified petroleum gas; gasoline, oils & petrochemicals; petroleum bulk stations, crude petroleum pipelines; plastics. Sells under trade name Phillips 66.

Subsidiaries
American Thermoplastics Corporation
Catalyst Resources Inc.-Catalyst Manufacture
GPM Gas Corporation
Phillips 66 Propane Company and Others.
It also has subsidiaries in South America, Europe, Africa, Australia, and Far East.

Ashland Oil, Inc.

Location, Chairman and Chief Executive Officer
Ashland, KY
John R. Hill

Economic Structure
Employees	33,400
Revenue	$8.50 billion

1918	Swiss Oil Corp. was incorporated in Kentucky.
1924	Owned 95 percent of capital stock of Ashland Refining Co.
1925	Acquired producing properties of Union Gas & Oil Co. located in Kentucky.
1930	Ashland Refining Co. organized Tri-State Refining Co. in West Virginia to acquire marketing facilities of Tri-State Refining Co. of Nevada.
1931	Organized Ashland Oil & Transportation Co. in Kentucky.
1936	Consolidation of Swiss Oil Corp. and Ashland Oil Refining Co. to form Ashland Oil & Refining Co.
1970	Present name adopted.
1979	Acquired 50 percent interest in Allemania Chemical Co., Plaquemine, LA.
	Formed Ashland Development Inc.
1980	Sold Levingston Shipbuilding Co. and its subsidiary, Levingston-Armadillo, Inc.
1981	Acquired the common stock of United States Filter Corporation.
1982	Acquired Scurlock Oil Co. of Houston, Texas.
1983	Acquired General Polymers National, Inc.
1986	Acquired Rapid Oil Change, Inc.
1989	Sold Ashland Technology Corp.
1991	Sold assets of Cleveland Tankers, Inc., Great Lakes shipping subsidiary.

1992   Acquired retail chemical distribution business from Union Oil of California.

Operations

Petroleum refining; transportation and marketing; retail gasoline marketing; motor oil and lubricants marketing; chemicals; coal; construction; oil and gas exploration and production.

Subsidiaries

Ashland Pipe Line Company
Allegheny Land Company
Arch Mineral Corporation
Ashland Chemical Inc.
Ashland Exploration Inc.
Ashland Oil & Transportation Co.
Ashland Overseas Investments Inc.
APAC Inc. in several states and numerous other subsidiaries.

Shell Oil Company

Location, President and Chief Executive Officer

Houston, TX
Frank R. Richardson

Economic Structure

Employees   31,637
Revenue $24.80 billion

1922   Incorporated as Shell Union Oil Corp.

1923   Acquired two-thirds outstanding common of Central Petroleum Co. (ME) and changed name to Wolverine Petroleum Corp.

1924   Sold stockholdings in Union Oil Co. of California.

1929   Acquired refinery, bulk terminal and distributing system formerly owned by New England Oil Refining Co.

Acquired stock of New Orleans Refining Co.

1932   Purchased half interest in properties in Kettleman North Dome Field in California from George F. Getty, Inc.

1939   Shell Oil Co. (CA) merged into another subsidiary, Shell Petroleum Corp.

1939
cont.
Atlantic Coast operating properties of Shell Union Oil Corp. were transferred to Shell Oil Co., Inc.

1945
Shell Chemical Corp. was formed.

1949
Shell Oil Co. was liquidated as subsidiary.

Adopted present name.

1959
Shell Oil Co. merged Shell Chemical Corp. and Shell Development Co. as divisions.

1961
Acquired Bishop Oil Co. and Canadian Bishop Oil Co. Ltd.

1962
Shell Oil Co. merged Shell Canadian Exploration Co.

1974
Shell Canadian Exploration Co. was terminated.

1977
Acquired Seaway Coal Co. of Cadiz, Ohio.

1979
Acquired Belridge Oil Co. of California.

1983
Sold its Northern Californian Geothermal Energy Project.

1985
Shell California Production Inc. acquired some oil and gas property from Marathon Oil Co.

Signed agreement with Atlantic Richfield for 400 Arco service stations.

1986
Sold agricultural chemical business to E. I. DuPont de Nemours & Co.

1987
Shell California Production Inc. merged into Shell Western E & P Inc.

1988
Shell Mining Inc. acquired Scallop Coal Corp. from Shell Energy Resources Inc.

Operations
Petroleum refining; crude petroleum pipelines; filling stations; gasoline; crude petroleum production; natural gas production, epoxy resins; chemicals.

Subsidiaries
Shell Pipe Line Corporation (DE)
Shell Energy Resources, Inc. (DE)

Pecten Arabian Company (DE)
Pecten Trading Company (DE)

Quaker State Corporation

Location, Chairman and Chief Executive Officer
Oil City, PA
Jack W. Corn

Economic Structure

Employees	4,400
Revenue	874.05 million

1931 — Incorporated as Quaker State Oil Refining Corp. to merge 19 companies engaged in refining and marketing Pennsylvania grade crude oil.

1987 — Adopted name of Quaker State Corporation.

1988 — Quaker State Minit-Lube, a subsidiary entered into a joint venture with a Canadian Quick Lube operation and founded Minit-Lube Ontario, Inc.

Operations
Manufacture and sale of brand name motor oils and other products and services; automotive aftermarket products, Quick-Lube service centers; coal operations, and insurance operations.

Subsidiaries
Heritage Insurance Group (CA)
Quaker State Minit-Lube Inc. (UT)
The Valley Camp Coal Co. (PA)
Truck-Lite Company, Inc. (NY)
Great Lakes Coal and Dock Co. (MN)
Quaker State, Inc. (Canada)
Valley Camp, Inc. (Canada)
McQuik's Oilube Inc.
Sturdivant Life Insurance Co.
Quaker State Oil Refining Corp. (PA)
QSE & P, Inc. (PA)
Quaker State Japan Co., Ltd. (Japan)

Getty Petroleum Corporation

Location, President and Chief Executive Officer
Jericho, NY
Leo Liebowitz

Economic Structure

Employees	2,500
Revenue	$1.26 billion

1971   Incorporated in Delaware as Power Test Corporation.

Acquired Leemilt's Petroleum, Inc.

Acquired Clay Oil Company and 14 gas stations in New York.

1972   Acquired Staten Island Gasoline, Inc. and its 16 gas stations in New York.

1978   Acquired gasoline marketing assets of Spiegel and Sons Oil Corp.

1985   Adopted present name.

Texaco took over all of Getty Oil Company's marketing and distribution activities in Northeast and Mid-Atlantic regions.

1986   Acquired Aero Oil Company.

1987   Acquired Kingston Oil Supply Corp.

1990   Purchased assets of Labone Petroleum Sales Corp.

1991   Agreement to lease on long-term basis 53 service stations and supply 73 retail service stations located in Massachusetts, Rhode Island, and New Hampshire.

Operations

Gasoline; middle distillates and other petroleum products, transportation, and marketing.

Subsidiaries

Aero Oil Company (PA)
Berkshire Construction Company (PA)
Getty Terminal Corporation (NY)
P.T. Petro Corporation (NY)
Power Test Petroleum Distributors (NY)

Westbury Small Business Corporation (NY)
Numerous Others.

Occidental Petroleum Corporation

Location, President and Chief Executive Officer
Los Angeles, CA
Ray R. Irani

Economic Structure

Employees	53,500
Revenue	$21.69 billion

1966–1985 California Company made a number of acquisitions for various amounts of cash and securities.

1986 Corporation was incorporated. Became parent of a company of the same name in California that was incorporated in California in 1920.

Acquired 100 percent ownership of MidCon.

Acquired Diamond Shamrock Chemicals Company.

1987 Acquired Shell Oil Company's vinyl chloride business.

1988 Acquired Cain Chemical Inc.

1991 Company sold its interest in the Antwerp Gas Terminal.

Sold its U.K. oil and gas subsidiaries.

Sold its interest in An Tai Bao coal mine in China.

Operations
Natural gas; crude oil; natural gas liquids; fertilizers and agricultural chemicals; coal; industrial chemicals; plastics and phosphate.

Subsidiaries
Occidental Petroleum Corporation (DE)
MidCon Marketing Corporation (DE)
Occidental Petroleum Investment Company (CA)
Occidental Oil and Gas Corporations (CA)

Organization of Petroleum Exporting Countries (OPEC)

1944 Arab League created to coordinate oil policy in the Middle East.

1951 Arab League established a Council of Arab oil experts.

1954 Arab League established a permanent office to exchange oil statistics and opinions between member states.

1959 Arab League held its initial Arab Petroleum Congress in Cairo. This meeting was attended by Arab country oil producers and also Venezuela and Iran. The Congress called for the permanent establishment of an Arab petroleum organization.

1960 August. International oil companies reduced the price of petroleum without consultation with the Middle East countries.

September. Middle East oil producing countries plus Iran and Venezuela established the Organization of Petroleum Exporting Countries to protect the unilateral price cutting actions of the international oil companies.

October. The international oil companies did not increase the price of a barrel of petroleum but informally told the director of the Arab League that in the future they would make no more cuts on posted prices before consulting and obtaining the approval of the oil-exporting countries.

1961 Secretariat established. The Secretariat is the body which carries out the executive function of OPEC.

Board of Governors. The Board represents the state of the relationship among the ministers.

OPEC Fund for International Development. Kuwait was the first to institutionalize its aid program: Kuwait Fund for Arab Economic Development.

Qatar joins OPEC.

Second Conference of OPEC meets in Caracas, Venezuela.

1962 Libya and Indonesia join OPEC.

1964   Economic Commission Board established as permanent, specialized organ of OPEC. Its functions were to examine the position of petroleum prices on a permanent basis; to study all economic and political factors that affect petroleum prices and their structure; to submit monthly reports on the price of petroleum; and to formulate and submit to the conference relevant recommendations.

1965   OPEC headquarters established in Vienna, Austria.

1970s   A number of OPEC countries established international aid funds: 1971, Abu Dhabi Fuel for Arab Economic Development; 1974, Iraq Fund for External Development; 1974, Saudi Fund for Development. Other OPEC countries established institutions that combined aid and investment.

In 1970 Libya began negotiations to increase the price of a barrel of oil. Occidental Oil Company was the first to come to an agreement with the Libyan government on September 1, 1970, with an agreement to increase the posted price of a barrel of crude of 40 degrees *API* from $2.23 to $2.53. The price was to increase by 2 cents a year for a period of five years. The company also agreed to pay a surtax of 3 percent per year on profits in satisfaction of government claims. As a reward the government increased Occidental's output from 464,500 to 700,000 barrels a day. Other companies could not resist the Libyan pressure for long and within two months of the Occidental agreement, all other companies had come to the same terms.

As soon as the Libyan-Occidental agreement was signed, the other oil companies sought and obtained antitrust clearance for U.S. oil companies to negotiate as a group with the producing countries. Fifteen oil companies sent OPEC an ultimatum outlining terms under which they would negotiate price increases.

1971   On February 15 the companies and OPEC nations signed the Tehran Agreement, which increased the posted price of oil by $0.35 at the Gulf terminal; and, from 1971 to 1975, there was to be an increase of 2.5 percent of the posted price plus 5 cents. Marketing allowances were also eliminated.

Libya did not agree with the Tehran Agreement and in March 1971 developed the Tripoli Agreement.

1971
cont.

In the Tripoli Agreement, Libya received 10 cents per barrel as a premium for its low-sulfur oil.

The earlier agreements were soon followed by the Geneva Currency Agreement. This agreement increased the price of oil by 8.49 percent. Future posted prices were tied to the U.S. dollar, the currency of posting prices, and other key currencies.

1973

In September 1973 the 35th OPEC Conference took the position that the level of posted prices determined by earlier agreements were "no longer compatible with the current and expected future trends of the world inflation as well as crude oil and product prices," and that oil companies were "reaping high increased profits."

Arab-Israeli War. Egyptian and Syrian governments' stated aim was to regain territories that were under Israeli occupation since 1967. The war increasingly created a new pricing attitude. This change in viewpoint resulted from the use of oil as a weapon in the war.

On October 15, 1973, the six Gulf oil-producing countries in Kuwait unilaterally increased the posted price of oil from $3.01 to $5.11 per barrel. An increase of 70 percent of the price that prevailed on October 1, 1973.

An oil embargo was declared on October 17, 1973, when the Arab oil-exporting nations decided to use the oil weapon in support of the Arab cause, because the United States continued to support Israel. The importing nations were divided into these categories:

1. Friendly nations—Those who opposed the Israeli occupation of Arab territories, mainly, the United Kingdom, France, Spain, and Eastern European countries.
2. Nonfriendly nations—Those who supported Israel, primarily the United States and the Netherlands.
3. Other nations were allowed to receive oil but at a reduced rate.

By November Arab oil production was down by more than 30 percent, from 20 million barrels daily to about 14 million barrels.

1973–1974   As a response to the cutbacks and low level of the posted price, the spot market prices reached astronomical highs. 1973–1974

To illustrate, Libyan oil went as high as $20 per barrel and Iranian oil to $17.35.

1974   March. The Arab oil embargo lifted and oil exports were resumed.

World oil supply increased and the price of oil fell by $0.40 to $3.00 per barrel.

OPEC countries were able to assert their power in all aspects of the oil industry within their borders. This included oil prices, participation, royalties, and income tax rates.

OPEC became the unchallenged power in determining the world price of crude oil.

1974–1976   Demand for OPEC oil decreased, which brought the posted price of oil down to $11.51. This price lasted until 1976.

1976   Ministers at the 55th OPEC Conference in June could not agree on an oil price. As a result, a two-tier price system developed when all OPEC nations except Saudi Arabia and the United Arab Emirates decided to increase the official price per barrel of oil from $11.51 to $12.70. This created confusion in the oil market.

1978–1979   In the 1978 Iran Revolution, oil production in the nation decreased from 6 million to 2.3 million barrels per day, creating uncertainty in the international oil market.

As a consequence, the spot market for oil increased to $23 for Arabian light and to over $26.00 for Libyan crudes.

1980   Oil prices continued to be unstable. In June the 57th OPEC Conference agreed to set the price for crude oil at $32.00 per barrel with a ceiling of $41.00.

This was the crest of the power of OPEC to establish the world price of petroleum. Power did not unite OPEC into a unified economic unit. Many countries developed their own philosophy of pricing.

1980
cont.

For example, Saudi Arabia and the United Arab Emirates had established a price per barrel, and then presented their deliberations to the OPEC Conference as a *fait accompli.*

OPEC, in order to maintain prices, attempted to establish production quotas. This system has not been successful because each nation wants to maintain its oil income.

In the 1980s other areas such as the North Sea, Soviet Union, and Mexico increased production. For example, in 1977 OPEC produced two-thirds of the total free world crude oil. By 1982 non-OPEC countries' production was a million barrels per day higher than OPEC output.

As a response to overproduction and decreases in consumption due to conservational practices, the price of oil fell from $29.00 per barrel for Arabian light oil in July 1980 to as low as $15.00 per barrel in the mid-1980s.

1990s

Because of the huge reserves of petroleum in the OPEC nations, these nations will continue to produce a major share of the world's petroleum needs.

1991

Iraq invades Kuwait—The United States comes to the assistance of Kuwait. Part of this decision was based on the need for the United States to have Kuwait's export of petroleum.

Profile of OPEC Nations

Indonesia

1893

Oil production began averaging 2,000 barrels per day.

1962

Member of OPEC.

1991

Crude oil production of 1,437,000 barrels per day.

1992

One national company, Indonesia National Petroleum Company (Pertamina), produces about 80,000 barrels daily. The rest of the production is produced by 16 foreign companies of which PT Caltex is most important.

Proved oil reserves set at 6.5 billion barrels.

Iran

1901   First Middle East oil concession granted to William D'Arcy.

1908   Oil discovered.

1931   First oil production.

1950s   When disputes between Iran and the Anglo-Iranian Oil Company were resolved, a consortium was established, with British Petroleum owning 50 percent of the shares; Royal Dutch Shell, 14 percent; Gulf, Exxon, Mobil, SOCAL, and Texaco, 7 percent each; CFP, 6 percent; and Iricon, 5 percent.

1950   Anglo-Iranian Oil Company nationalized by Dr. Mossadeq.

1960   Member of OPEC.

1973   Consortium became a service company.

1980   Iranian government canceled all service contracts, participation agreements, exploration, and production.

Nationalized industry served by National Iranian Oil Company and the Continental Shelf Oil Company of Iran.

1991   Crude oil production of 3,167,000 barrels per day.

1992   Proved reserves estimated at 92.8 billion barrels.

Venezuela

1917   Oil production began.

Venezuela's oil industry owned by Gulf and Royal Dutch Shell.

1960   Member of OPEC.

1975   Oil industry nationalized in December 1975. It is now run by four companies: Lagoven, Maraven, Meneven and Corpoven.

1991   Crude oil production of 2,329,000 barrels per day.

1992   Proven reserves estimated at 59 billion barrels.

Iraq

1923   Oil production began, producing 2,700 barrels a day.

1960   Member of OPEC.

1972   Iraqi Petroleum Company was nationalized in June 1972. An arbitrated settlement was agreed upon in June 1973, transferring to the Iraqi government the assets of the Mosul Petroleum Company.

1975   Basra Petroleum Company was nationalized.

1977   The Iraqi government established two companies, the Northern Petroleum Company and the Southern Petroleum Company, to run petroleum operations in the country.

1992   Estimated recoverable oil reserves of 65 billion barrels.

Saudi Arabia

1933   Standard Oil of California (SOCAL) secures oil concession.

1934   Oil discovered by Standard Oil of California.

1935   Arabian American Oil Company formed as a partnership between SOCAL, Texaco, Gulf, and Mobil.

1980s   Arabian American Oil Company is 20 percent owned by the government and 80 percent by foreign companies.

1980   ARAMCO nationalized. Government ownership of this company came as a result of a participation agreement which gave ARAMCO's previous owners— SOCAL, Texaco, Gulf, and Mobil—compensation based on an "up-dated book value," plus a service fee "around 27 cents per barrel," and access to large quantities of crude oil.

1991   Crude oil production of 8,079,000 barrels per day.

1992   Proven estimated reserves at 257 billion barrels, the largest of the OPEC nations.

Kuwait

1938	Oil discovered.
	Kuwait Oil Company owned by British Petroleum and Gulf Oil.
1946	Oil production began with an average output of 162,000 barrels a day.
1970s	Kuwait Oil Company is progressively nationalized. Financial compensation and access to Kuwait oil given to British Petroleum and Gulf Oil.
1977	American Oil Company nationalized.
1990	Crude oil production of 1,073,000 barrels per day.
1991	Crude oil production of 127,000 barrels per day.
1992	Estimated oil reserves about 92 billion barrels.

Qatar

1949	Oil production began with an output of about 2,000 barrels a day.
1961	Member of OPEC.
1977	Qatar Petroleum Company became the sole petroleum-producing company after nationalizing Shell's 40 percent equity in the Shell Company of Qatar.
1991	Crude oil production of 378,000 barrels per day.
1992	Proved oil reserves at 3.7 billion barrels.

Algeria

1958	Oil production began, producing 8,800 barrels a day.
1969	Member of OPEC.
1991	Crude oil production of 800,000 barrels per day.
1992	The most important oil-producing company is National Company Sonatrack, which is 100 percent state-owned.

1992
cont.

The Algerian government has a 51 percent equity in four other companies—Total Algerian, Medaloil, Hispanoil, and Braspetro.

Estimated recoverable oil reserves of 9.2 billion barrels.

Nigeria

1958

Oil production began in 1958 with an output of 5,100 barrels a day.

1971

Member of OPEC.

1991

Crude oil production of 1,885,000 barrels per day.

1992

Nigerian National Petroleum Company owns 80 percent of the major producing company, Nigeria Shell/NNPC. The Nigerian National Petroleum Company also owns 60 percent of 5 foreign companies—Gulf, Mobil, Agip, Elf, and Texaco. Three other companies are entirely foreign owned— Pan Ocean, Tenneco, and Phillips.

Estimated proved reserves of 17.9 billion barrels.

Libya

1959

Petroleum discovered.

1961

Oil production started averaging 18,200 barrels a day.

1962

Member of OPEC.

1991

Crude oil production of 1,413,000 barrels per day.

1992

The Libyan oil industry is not fully owned by the government. Of the 10 companies operating in the nation only two (Arabian Gulf and Umm Gawaby) are state owned. The government's share in other companies varies. Its share in Oasis Oil Company is 59.17 percent, 50 percent in Agip, 51 percent in Mobil, Occidental, and Esso Standard, and it has 63.5 percent in Esso Sirte. Wintershall, a small German company, is 100 percent foreign owned.

Proved oil reserves are 22.8 billion barrels.

United Arab Emirates

1962   Oil production began with an average rate of 14,200 barrels per day.

1967   Member of OPEC.

1991   Crude oil production of 2,388,000 barrels per day.

1992   Seven oil companies produce oil in the United Arab Emirates. The importance of foreign ownership varies. Five companies in Abu Dhabi, of which the government owns 60 percent of Adma, Adco, and 25 percent of ADNOC. Total and Amerada Hess are totally owned by foreign companies. In Dubai, one company, Dubai Petroleum Company, is 100 percent owned by the government, and in Sharjah, Buttes is a foreign company.

Natural Gas

Table 2.13
United States Natural Gas Reserves

Year	Proved Reserves (Dec. 31)	Reserve Revisions, Extensions and Discoveries During Year	Net Change in Underg round Storage	Production	Proved Reserves at Year's End
1960	261,170,431	13,893,978	281,272	13,019,356	262,326,326
1965	281,251,454	21,319,279	150,483	16,252,293	286,468,923
1970	275,108,835	37,196,359	402,018	21,960,804	290,746,408
1975	237,132,497	10,480,688	302,561	19,718,520	228,200,176
1980	200,997,000	16,723,000	NA	18,699,000	199,021,000
1985	197,463,000	11,891,000	NA	15,985,000	193,369,000
1990	167,116,000	19,463,000	NA	17,233,000	169,346,000

Source: *Basic Petroleum Data Book.* Annual. Washington, D.C.: American Petroleum Institute.

Table 2.14
Natural Gas Production by States
(Trillion Cubic Feet)

1960 United States	15.09
Texas	6.96
Louisiana	3.31
Oklahoma	1.13
Other	3.68
1970 United States	23.79
Texas	9.40
Louisiana	8.08
Oklahoma	1.81
Other	4.50
1980 United States	21.87
Texas	7.66
Louisiana	7.01
Oklahoma	2.02
Other	5.19
1990 United States	21.24
Texas	6.93
Louisiana	5.27
Oklahoma	2.18
Other	6.87

Source: *Natural Gas Annual.* Washington, D.C.: Energy Information Administration, Office of Oil and Gas, U.S. Department of Energy.

Table 2.15
Gas Well Productivity

Year	Gross Withdrawal from Gas Wells (Trillion Cubic Feet)	Producing Wells (000)	Average Productivity (Thousands cubic feet per day)
1960	10.85	91	326.7
1970	18.59	117	433.6
1980	17.57	182	263.8
1990	15.88	265	164.4

Source: *Natural Gas Annual.* Washington, D.C.: Energy Information Administration, Office of Oil and Gas, U.S. Department of Energy. *Natural Gas Monthly.* Washington, D.C.: Energy Information Administration, Office of Oil and Gas, U.S. Department of Energy.

Table 2.16
U.S. Imports of Natural Gas by Source
(Million Cubic Feet)

Year	Canada	Mexico	Algeria	Total Imports	U.S. Natural Gas Consumption
1970	780	41	<500	821	22,046
1975	948	0	5	953	19,538
1980	797	102	86	985	19,877
1990	1,448	0	84	1,523	18,714

Source: *Basic Petroleum Data Book,* Annual. Washington, D.C: American Petroleum Institute.

Coal

Table 2.17
Estimated Original Coal Reserves
(Thousands of Tons)

Year	Bituminous	Subbituminous	Lignite	Anthracite	Total
1950	1,280,735,000	468,544,160	711,693,233	23,663,700	2,484,636,637
1960	808,420,000	437,742,000	447,966,000	25,836,000	1,719,964,000
1970	729,545,000	429,083,000	448,083,000	23,717,000	1,624,509,000
1974	747,357,000	485,766,000	478,134,000	19,662,000	1,730,919,000

Source: *Minerals Yearbook.* Annual. Washington, D.C.: Bureau of Mines, U.S. Department of the Interior.

Table 2.18
Coal Production by States
(Thousands of Tons)

1950 United States	516,311
West Virginia	145,565
Pennsylvania	102,500
Kentucky	77,900
Illinois	55,346
Ohio	<u>36,936</u>
Total (81 percent of U.S. total)	418,247
1960 United States	415,512
West Virginia	118,944
Kentucky	66,847
Pennsylvania	65,425
Illinois	45,977
Ohio	<u>33,957</u>
Total (79 percent of U.S. total)	331,150
1970 United States	602,932
West Virginia	144,072
Kentucky	125,505
Pennsylvania	80,491
Illinois	65,119
Ohio	<u>55,351</u>
Total (78 percent of U.S. total)	470,538
1980 United States	825,673
Kentucky	145,986
West Virginia	120,349
Wyoming	94,968
Pennsylvania	92,951
Illinois	62,361
Ohio	<u>39,178</u>
Total (67 percent of U.S. total)	555,793
1990 United States	1,026,307
Wyoming	184,249
Kentucky	172,546
West Virginia	168,720
Pennsylvania	69,600
Illinois	<u>60,393</u>
Total (63 percent of U.S. total)	652,382

Sources: *Minerals Yearbook*. Annual. Washington, D.C.: Bureau of Mines, U.S. Department of the Interior; *Coal Production*. Annual. Washington, D.C.: U.S. Department of Energy.

Table 2.19
Coal Production by Rank
(Thousands of Tons)

1980 United States		825,673
Bituminous		618,363
Kentucky	145,986	
West Virginia	120,349	
Pennsylvania	86,994	
Illinois	62,361	
Ohio	39,178	
Subbituminous		153,932
Wyoming	93,800	
Montana	29,578	
New Mexico	17,798	
Lignite		47,375
Texas	30,079	
Anthracite		6,003
Pennsylvania	5,957	
1990 United States		1,026,307
Bituminous		690,817
Kentucky	172,546	
West Virginia	168,720	
Pennsylvania	66,474	
Illinois	60,393	
Indiana	35,886	
Subbituminous		244,274
Wyoming	182,425	
Montana	37,266	
Lignite		88,090
Texas	55,400	
North Dakota	29,213	
Anthracite		3,126
Pennsylvania	3,126	

Source: *Coal Production.* Annual. Washington, D.C.: U.S. Department of Energy.

Table 2.20
Coal Production by Type of Mining
(Thousands of Tons)

1970 United States		602,932
Underground		338,788
West Virginia	116,414	
Kentucky	62,610	
Pennsylvania	55,382	
Illinois	32,090	
Virginia	28,018	
Surface		244,117
Kentucky	52,836	
Ohio	35,818	
Illinois	33,026	
Pennsylvania	24,447	
Indiana	20,169	
1980 United States		825,673
Underground		329,474
West Virginia	95,795	
Kentucky	74,954	
Pennsylvania	41,448	
Illinois	34,960	
Virginia	31,449	
Surface		486,719
Wyoming	94,033	
Kentucky	71,033	
Pennsylvania	51,503	
Indiana	30,297	
Texas	30,180	
Montana	29,948	
Illinois	27,401	
1990 United States		1,026,307
Underground		423,556
West Virginia	123,306	
Kentucky	105,290	
Illinois	41,671	
Pennsylvania	40,104	
Virginia	39,100	
Surface		604,529
Wyoming	182,527	
Kentucky	68,032	
Texas	55,755	
West Virginia	45,898	
Montana	37,616	

Source: *Minerals Yearbook.* Annual. Washington, D.C.: Bureau of Mines, U.S. Department of the Interior; *Coal Production,* Annual. Washington, D.C.: U.S. Department of Energy.

Table 2.21
Major Coal Companies 1989

Rank	Coal Company	Total Production (Tons)	Percent of U.S. Total Production
1	Peabody Holding Co.	78,819	8.17
2	E. I. Du Pont de Nemours	50,787	5.26
3	Texas Utilities Mining Co.	29,917	3.10
4	North American Coal Corp.	22,223	2.30
5	Sun Oil Co.	21,809	2.26
6	Montana Power Co.	21,797	2.25
7	Shell Mining Co.	21,641	2.24
8	Arch Mineral Corp.	21,078	2.17
9	Occidental Petroleum Corp.	17,842	1.84
10	NERCO Coal Corp.	17,067	1.76
11	Cyprus Coal Co.	14,095	1.45
12	Bethlehem Steel Corp.	10,673	1.10
13	A. T. Massey Coal Co., Inc.	8,406	0.87
14	Mobil Corp.	8,369	0.86
15	Sunland Mining Co.	7,178	0.73
16	Pacificorp Electric Operations	6,840	0.70
17	Ashland Coal Co.	6,453	0.66
18	Addington Resources, Inc.	4,188	0.42
19	Texas Municipal Power Agency	3,482	0.35
20	San Miguel Electric Corp., Inc.	2,984	0.30

Source: *Coal Production Report.* Form E1A-7A. Washington, D.C.: Energy Information Administration, U.S. Department of Energy.

Nuclear Energy

Table 2.22
Uranium Supply, Enrichment and Discharged Commercial Reactor Fuel

	Unit	1970	1975	1980	1985	1990
Uranium Concentration						
Production	Mil. lb.	25.81	23.20	43.70	11.31	8.90
Exports	Mil. lb.	4.20	1.00	5.80	5.30	2.00
Imports	Mil. lb.	—	1.40	3.60	11.70	23.70
Delivered Price	Dol./lb.	NA	10.50	26.00	31.43	15.70
Enrichment[3]						
Enriched Product[1]	Mil. SWU[2]	5.10	9.92	10.69	10.2	10.2
For Domestic Consumers	Mil. SWU[2]	3.74	4.36	6.89	6.0	6.8
For Foreign Consumers	Mil. SWU[2]	1.36	5.56	3.80	4.2	3.4
Sales	Mil. dol.	NA	376	1,379	1,403	1,148
Discharged Commercial Reactor Fuel[4]						
Annual Discharge	Metric tons	82	499	1,193	1,330	NA
Inventory: Yearend[5]	Metric tons	118	1,538	6,434	12,481	NA

[1]Based on sales.
[2]SWU (Separate Work Units) standard measure of enrichment services is based on operating tails assay in effect at the time the enriched product was placed in inventory.
[3]Beginning 1984, represents fiscal years.
[4]Uranium content source, Nuclear Assurance Corporation, Atlanta, GA.
[5]Reprocessed fuel not included as inventory.

Source: *Annual Energy Review.* Annual. Washington, D.C.: Energy Information Administration, U.S. Department of Energy.

Table 2.23
U.S. Commercial Nuclear Power Generation

Year	Number of Reactors	Electricity Generated (Bil, KWH)	Gross Capacity (1,000 KW)
1970	15	23.2	5,211
1980	74	265.2	56,529
1985	89	399.1	80,327
1990	110	555.2	103,643

Source: *Statistical Abstract of the United States.* Annual. Washington, D.C.:Bureau of the Census, U.S. Department of Commerce.

Table 2.24
Nuclear Power Plants

	Number of Units	Net Generation Total (Mil, KW)	Net Generation Percent of Total
United States	111	576,784	20.6
Northeast	28	142,585	33.6
New England	9	37,404	39.8
Maine	1	4,861	53.6
New Hampshire	1	4,081	37.7
Vermont	1	3,616	72.4
Massachusetts	2	5,070	13.9
Connecticut	4	19,776	61.5
Middle Atlantic	19	105,181	31.8
New York	6	23,623	18.4
New Jersey	4	23,770	65.1
Pennsylvania	9	57,787	34.9
Midwest	31	153,923	21.9
East North Central	23	115,388	23.8
Ohio	2	10,664	8.4
Illinois	13	71,887	56.6
Michigan	5	21,610	24.3
Wisconsin	3	11,226	24.6
West North Central	8	38,535	17.8
Minnesota	3	12,139	30.8
Iowa	1	3,012	10.4
Missouri	1	7,998	13.6
Nebraska	2	7,511	34.7
Kansas	1	7,874	23.3

(table cont'd)

South	42	215,169	18.6
South Atlantic	27	140,354	26.3
Maryland	2	1,251	4.0
Virginia	4	23,820	50.5
North Carolina	5	25,905	32.5
South Carolina	7	42,881	61.9
Georgia	4	24,797	25.4
Florida	5	21,699	17.6
East South Central	8	33,477	13.6
Tennessee	2	14,003	19.0
Alabama	5	12,052	15.8
Mississippi	1	7,422	32.3
West South Central	7	41,338	11.1
Arkansas	2	11,282	30.4
Louisiana	2	14,197	24.7
Texas	3	15,589	6.8
West	10	65,110	12.4
Mountain	3	20,598	8.3
Arizona	3	20,598	33.1
Pacific	7	44,512	16.9
Washington	1	5,742	5.7
Oregon	1	6,074	12.4
California	5	32,696	28.5

Source: *Electric Power Monthly,* Washington, D.C.: Office of Coal and Electric Power Statistics, U.S. Department of Energy.

Renewable Energy

Table 2.25
Renewable Energy Consumption Estimates
(Quadrillion Btu)

1990 U.S. Total Energy Consumption	84.99	
Total renewable energy consumption estimates		6.70
Reported renewable energy consumption		3.15
Hydroelectric Power	2.94	
Geothermal energy at Electric Utilities	0.18	
Wood and Waste Energy at Electric Utilities	0.02	
Wind Energy at Electric Utilities	< 50 billion	
Additional renewable energy consumption estimates		3.55
Biofuels	3.23	
Geothermal Energy (nonelectric utilities)	0.16	
Solar Energy (nonelectric utilities)	0.08	
Wind Energy (nonelectric utilities)	0.04	
Hydroelectric Power (additional industrial use)	0.04	

Source: *Annual Energy Review.* Washington, D.C.: Energy Information Administration, U.S. Department of Energy.

Table 2.26
Water Power Developed and Undeveloped
(In Millions of Kilowatts)

1950	Developed	Undeveloped
United States	18.7	87.6
New England	1.2	3.3
Middle Atlantic	1.7	6.6
East North Central	0.9	2.3
West North Central	0.6	5.8
South Atlantic	2.8	8.2
East South Central	2.7	4.7
West South Central	0.5	3.6
Mountain	2.3	23.4
Pacific	6.0	29.8

1960	Developed	Undeveloped
United States	33.2	114.2
New England	1.5	2.9
Middle Atlantic	2.5	7.6
East North Central	0.9	3.0
West North Central	1.6	6.4
South Atlantic	3.8	8.4
East South Central	3.8	4.6
West South Central	0.9	3.9
Mountain	4.6	23.6
Pacific	13.6	53.8

1970	Developed	Undeveloped
United States	52.0	128.0
New England	1.5	3.3
Middle Atlantic	4.3	4.5
East North Central	0.9	1.6
West North Central	2.7	4.4
South Atlantic	5.3	9.6
East South Central	5.2	3.8
West South Central	1.9	3.3
Mountain	6.2	26.7
Pacific	23.9	70.9

(table cont'd)

1980		Developed	Undeveloped
	United States	64.4	129.9
	New England	1.5	4.7
	Middle Atlantic	4.3	5.1
	East North Central	0.9	2.0
	West North Central	2.8	3.4
	South Atlantic	5.9	9.6
	East South Central	5.6	3.3
	West South Central	2.3	4.7
	Mountain	7.4	34.2
	Pacific	33.7	62.9
1990		Developed	Undeveloped
	United States	73.0	73.9
	New England	1.9	4.4
	Middle Atlantic	4.9	5.1
	East North Central	1.1	1.7
	West North Central	3.1	3.1
	South Atlantic	6.7	7.0
	East South Central	5.9	2.4
	West South Central	2.7	4.6
	Mountain	9.2	19.4
	Pacific	37.5	26.2

Source: *Statistical Abstract of the United States.* Annual. Washington, D.C.: Bureau of the Census, U.S. Department of Commerce.

Table 2.27
Solar Collector Shipments

Year	Number of Manufacturers	Total Shipments 1	Collector Type		End Use			Market Sector		
			Low Temp.	Max. Temp.	Pool Heating	Hot Water	Space Heating	Residential	Commercial	Industrial
1975	131	3,743	3,026	717	NA	NA	NA	NA	NA	NA
1980	233	19,398	12,233	7,165	12,029	4,790	1,688	16,077	2,417	488
1984	225	17,191	4,479	11,939	4,427	8,930	2,370	13,980	2,091	289
1989	44	11,482	4,283	1,989	4,688	1,374	205	5,804	424	42

Source: *Solar Collector Manufacturing Activity.* Annual. Washington, D.C.: U.S. Energy Information Administration, U.S. Department of Energy.

Table 2.28
Wood Energy Consumption
(In Trillions of Btu)

1980 United States   2,483

Region

Northeast	386
Midwest	329
South	1,380
West	388

Sector

Residential	859
Industrial	1,600
Commercial	21
Electric Utilities	4

1989 United States   2,487

Region

Northeast	413
Midwest	527
South	1,109
West	438

Sector

Residential	859
Industrial	1,600
Commercial	21
Electric Utilities	13

Source: *Statistical Abstract of the United States.* Annual. Washington, D.C.: Bureau of the Census, U.S. Department of Commerce.

Table 2.29
Households That Burn Wood

	Unit	1980	1987
Number of households	Millions	21.6	22.5
Percent of all households	Percent	26.4	24.8
Number of cords burned	Millions	42.7	42.6
Average number per household	Number	2.0	1.9
Median number per household	Number	0.7	0.7
Wood Energy Consumption	Tril. Btu	854.0	853.0

Table 2.30
Households That Burn Wood as Main Heating Fuel

	Unit	1980	1987
Number of households	Millions	4.7	5.0
Percent of all households	Percent	5.8	5.6
Number of cords burned	Millions	22.4	23.5
Average number per household	Number	4.7	4.7
Median number per household	Number	3.3	4.0
Wood Energy Consumption	Tril. Btu	448	470

Source: *Annual Energy Review.* Washington, D.C.: Energy Information Administration, Office of Energy Markets and End Use, U.S. Department of Energy.

Chronology of Legislation

Basic Energy Legislation

1974   Energy Supply and Environmental Coordination Act

1975   Energy Policy and Conservation Act

1978   National Energy Conservation Policy Act

Public Utilities Regulating Policies Act

1982   Energy Emergency Preparedness Act

1985   Energy Policy and Conservation Amendment Act

1990   Energy Policy and Conservation Act Extension Amendment

Energy Policy and Conservation Act Short-term Extension Amendment

Energy Efficiency Legislation

1974   Federal Energy Administration Act

Energy Reorganization Act

1976	Energy Conservation and Production Act
	Energy Conservation Standards for New Buildings
	Energy Conservation Standard for Existing Buildings
1978	Energy Tax Act
1980	Crude Oil Windfall Profit Tax
1987	National Appliance Energy Conservation Act
1988	Federal Energy Management Improvement Act

Energy and the Environment

Clean Air Acts

1955	Clean Air Act, Chapter 360, July 14, 1955
1970	Clean Air Act, Public Law 91-604, December 31, 1970
1977	Clean Air Act, Public Law 95-190, November 16, 1977
1990	Clean Air Act, Public Law 101-549, November 15, 1990

Coal Mining

1977	Surface Mining Control and Reclamation Act, Public Law 95-87, August 3, 1977

Atomic Energy

1946	Atomic Energy Act of 1946
1954	Atomic Energy Act of 1954
1955	Atomic Energy Community Act
1957	Atomic Energy Damage Act

Natural Gas Legislation

1938	Natural Gas Act

1968   Natural Gas Pipeline Safety Act

1978   Natural Gas Policy Act

1988   Natural Gas Act

1989   Natural Gas Wellhead Decontrol Act

Renewable Energy

Basic Legislation

1978   Renewable Resources Extension Act

1980   Energy Security Act, Public Law 96-294, June 30, 1980

                 Geothermal Energy Act of 1980

                 Solar Energy and Conservation Act of 1980

                 Solar Energy and Energy Conservation Bank Act of 1980

                 Biomass Energy and Alcohol Fuels Act of 1980

                 Renewable Energy Resources Act of 1980

1989   Renewable Energy and Efficiency Technology
                 Competitiveness Act of 1989, Public Law 101-218, December
                 11, 1989

1990   Solar, Wind, Waste, and Geothermal Power Production
                 Incentive Act of 1990, Public Law 101-575, November 15, 1990

Hydroelectric Power

1920   Federal Water Power Act, Chapter 285, June 10, 1920

1970   Energy and Water Development Appropriation Act of 1970,
                 Public Law 96-69, September 25, 1970

Geothermal Energy

1970   Geothermal Steam Act of 1970, Public Law 91-581,
                 December 24, 1970

Solar Energy

1974   Solar Energy Research Development and Demonstration Act of 1974, Public Law 93-473, October 26, 1974

1974   Solar Heating and Cooling Demonstration Act of 1974, Public Law 93-409

1975   Solar Photovoltaic Energy Research Development and Demonstration Act, Public Law 95-590

Wind Energy

1980   Wind Energy System Act, Public Law 96-345, September 8, 1980

3

Laws and Regulations

RECOGNITION THAT THE UNITED STATES IS NOT SELF-SUFFICIENT in its domestic requirements for petroleum has triggered the need to investigate the total energy requirements for the nation. As a response Congress has passed a wide range of laws treating such topics as conservation of energy, substitution of plentiful energy for atomic energy, scarce energy, and renewable energy.

Basic Energy Legislation

Energy Supply and Environmental Coordination Acts

June 22, 1974, Public Law 93-319, 88 Statute 246

December 22, 1975, Public Law 94-163, 89 Statute 875

July 21, 1977, Public Law 95-70, 91 Statute 277

November 9, 1978, Public Law 95-620, 92 Statute 3346

Energy Supply and Environmental Coordination Act of 1974 (Public Law 93-319, 88 Statute 246)

The purpose of this legislation was to grant specific authority to increase the use of coal resources so as to increase energy supplies from domestic sources, to obtain information about energy supplies within the nation, and to permit certain adjustments of environmental requirements so that the nation's essential energy needs could be met in a manner consistent with the national commitment to protect and improve the environment.

This act directs the steps to be taken to make more effective use of the nation's coal resources, authorizes the Federal Energy Administration to determine the nation's energy supply situation, and permits narrowly defined and limited variances from certain specific Clean Air Act requirements in order to ensure a sufficient supply of energy and still maintain environmental protection.

This act provided for a number of specific measures, which include:

1. Coal Conversion Allocation
   a. The Federal Energy Administration may prohibit any major fuel-burning installation, other than a power plant, from burning natural gas or petroleum products as its primary energy source if the administration finds that the installation has the capability and necessary plant equipment to burn coal.
   b. The Federal Energy Administration may require that any plant in the early planning process (other than a combustion gas turbine or combined cycle unit) be designed and constructed so as to be capable of using coal as its primary energy source. No plant will be required to do this until the administration determines that: (1) to do so is likely to result in an impairment of reliability or adequacy of services, or (2) an adequate and reliable supply of coal is not expected to be available.
2. Suspension Authority
   This act gives the Federal Energy Administration authority to temporarily suspend any stationary source fuel or emission limitation if fuels are available that will comply with the Clean Air Act.
3. Motor Vehicle Emissions
   The standards for motor vehicle emissions were extended. However, a request could be made by the automobile manufacturers to suspend enforcement for one year.

4. Protection of Public Health and Environment
   This section contains provisions to ensure that the use
   of certain fuels will not impair health. For example, the
   act includes measures to ensure that available low-sulfur
   fuel will be distributed on a priority basis to those areas
   of the United States defined by the Environmental
   Protection Agency as requiring low-sulfur fuel to avoid
   or minimize adverse impact on public health.
5. Energy Conservation Study
   This section contains provisions for conducting studies
   on potential methods of energy conservation including:
   a. Energy conservation potential of restricting exports of
      fuels or energy-intensive products or goods.
   b. Alternative requirements, incentives, or disincentives
      for increasing industrial recycling or resource re-
      covery in order to reduce energy demand, including
      economic costs and fuel consumption tradeoffs.
   c. Measure for incentives or disincentives to increase
      efficiency of the industrial use of energy.
6. Fuel Economy Study
   This section provided for the EPA and Department of
   Transportation to conduct a joint study on the feasibility
   of establishing a 20 percent fuel economy improvement
   standard for 1980 and later motor vehicles.
7. Reporting of Energy Information
   The Federal Energy Administration is to collect information on energy to assist in the formulation of an
   energy policy.

Energy Policy and Conservation Acts (EPCA)

December 22, 1975, Public Law 94-163, 89 Statute 871

April 5, 1976, Public Law 94-258, 90 Statute 305

August 15, 1976, Public Law 94-385, 90 Statute 1140, 1142,
1158, 1160

July 21, 1977, Public Law 95-70, 91 Statute 276, 277

November 9, 1978, Public Law 95-619, 92 Statute 3239, 3248,
3257, 3259, 3262, 3263, 3265, 3267, 3273, 3275, 3282, 3283,
3285, 3288

November 9, 1978, Public Law 95-620, 92 Statute 3347

June 30, 1979, Public Law 96-30, 93 Statute 80

September 29, 1979, Public Law 96-72, 93 Statute 535

October 31, 1979, Public Law 96-94, 93 Statute 720

November 5, 1979, Public Law 96-102, 93 Statute 751–756

November 30, 1979, Public Law 96-133, 93 Statute 1053

June 30, 1980, Public Law 96-294, 94 Statute 775,776

October 19, 1980, Public Law 96-470, 94 Statute 2242

March 13, 1981, Public Law 97-5, 95 Statute 7

August 13, 1981, Public Law 97-35, 95 Statute 618–620

September 30, 1981, Public Law 97-50, 95 Statute 957

April 1, 1982, Public Law 97-163, 96 Statute 24

June 1, 1982, Public Law 97-190, 96 Statute 106

June 19, 1982, Public Law 97-217, 96 Statute 196

August 3, 1982, Public Law 97-229, 96 Statute 248,
250–252

March 20, 1984, Public Law 98-239, 98 Statute 93

July 18, 1984, Public Law 98-370, 98 Statute 1211

October 5, 1984, Public Law 98-454, 98 Statute 1736

July 2, 1985, Public Law 99-58, 99 Statute 102–105

August 15, 1985, Public Law 99-88, 99 Statute 342

April 7, 1986, Public Law 99-272, 100 Statute 141

October 21, 1986, Public Law 99-509, 100 Statute
1888–1890

March 17, 1987, Public Law 100-12, 101 Statute 103 to 124

June 28, 1988, Public Law 100-357, 102 Statute 671

July 19, 1988, Public Law 100-373, 102 Statute 878

October 14, 1988, Public Law 100-494, 102 Statute
2442–2447

November 5, 1988, Public Law 100-615, 102 Statute 3189

June 30, 1989, Public Law 101-46, 103 Statute 132

December 11, 1989, Public Law 101-218, 103 Statute 1867

March 31, 1990, Public Law 101-262, 104 Statute 124

August 10, 1990, Public Law 101-360, 104 Statute 421

September 15, 1990, Public Law 101-383, 104 Statute 727–735

October 18, 1990, Public Law 101-440, 104 Statute 1006–1015

November 14, 1990, Public Law 101-548, 104 Statute 2398

Energy Policy and Conservation Amendment Acts

July 2, 1985, Public Law 99-58, 99 Statute 102

September 15, 1990, Public Law 101-383, 104 Statute 727

Energy Policy and Conservation Act Extension Amendment

March 31, 1990, Public Law 101-262, 104 Statute 124

Energy Policy and Conservation Act Short-Term Extension Amendment

August 10, 1990, Public Law 101-360, 104 Statute 421

Energy Policy and Conservation Act of 1975 (Public Law 94-163, 89 Statute 871)

This legislation is directed to the attainment of the collective goals of increasing domestic supply, conserving and managing energy demand, and establishing standby programs for minimizing this nation's vulnerability to major interruptions in the supply of petroleum imports. The energy policies which its terms define are tempered by current economic concerns and the compelling need to restore economic health.

The specific purposes of the act were to:

1. Grant specific standby authority to the president, subject to congressional review, to impose rationing, reduce demand for energy through the implementation of

energy conservation plans, and fulfill obligations of the United States under the international energy program

2. Provide for a Strategic Petroleum Reserve capable of reducing the impact of severe energy supply interruptions
3. Increase the supply of fossil fuels in the United States through price incentives and production requirements
4. Conserve energy supplies through energy conservation programs, and, where necessary, the regulation of certain energy uses
5. Provide for improved energy efficiency of motor vehicles, major appliances, and certain other consumer products
6. Reduce the demand for petroleum products and natural gas through programs designed to provide greater availability and use of the abundant coal reserves in the United States
7. Provide a means for verification of energy data to ensure the reliability of energy data

The enactment of the Energy Policy and Conservation Act provided the first major national move toward the establishment of an effective federal conservation effort. The EPCA directs the Federal Energy Administration to establish several specific conservation programs, including:

1. A program to provide technical and financial assistance to states
2. The establishment of energy efficient targets for the ten most energy consumptive industries, and monitoring the progress toward achieving these goals
3. The setting of efficiency targets for major home appliances and the development of procedures for determining energy efficiency
4. The development of a responsible public education program

Title I of the act dealt with Matters Related to Domestic Supply Available. A number of actions were possible to ensure a domestic supply of energy. They included:

1. Guaranteed loans for the development of underground coal mines and oil reserves. A person was eligible for a loan if such person:
   a. Did not produce more than 1 million tons of coal in the preceding calendar year.
   b. Did not produce more than 300,000 barrels of crude oil and did not own an oil refinery.
   c. Did not gross revenues in excess of $100,000.
2. The president may restrict exports of:
   a. Coal, petroleum products, natural gas, or petrochemical feedstock.
   b. Supplies of materials or equipment which he determines to be necessary to maintain for further exploration, production, refining, or transportation of energy supplies.
   c. Materials for the construction or maintenance of energy facilities within the United States.
3. Production of oil and natural gas at the maximum efficient rate and, when necessary, at a temporary emergency production ratio. The maximum official rate means a sustained production without loss of ultimate recovery of crude oil or natural gas.

Title I also includes the development of a Strategic Petroleum Reserve. The Strategic Petroleum Reserve was created because:

> Congress finds that the storage of substantial quantities of petroleum products will diminish the vulnerability of the United States to the effect of a severe energy supply interruption and provides limited protection from the short-term consequences of interruptions in supplies of petroleum products.

This statement established the policy of the United States to:

> Provide for the creation of a Strategic Petroleum Reserve for the storage of up to one billion barrels of petroleum products but not less than 150 million barrels of petroleum products ... It is further declared to be the policy of the United States to provide for the creation of an Early Storage Reserve, as part of the reserve for the purpose of providing limited protection from the impact of near-term disruption of supplies of petroleum products.

The Strategic Petroleum Reserve Plan includes:

1. A comprehensive environmental assessment
2. A description of the type and proposed location of each storage facility
3. The proximity of each storage facility to related facilities
4. Estimates of the volume and type of petroleum products proposed to be stored
5. A projection of the aggregate size of the reserve
6. An estimate of the cost of the reserve

Title II, Part A, of the act was devoted to Standby Energy Authorities. The implementation of this title included the establishment of an energy conservation contingency plan and a rationing contingency plan. Part B of the title considers international oil allocation and an international oil voluntary agreement.

Title V considers energy efficiency. Standards were established for each vehicle as to the number of miles per gallon. Part B provides an energy conservation program for consumer products other than automobiles. An energy efficiency standard was established which presented a minimum level of energy efficiency for a covered product, determined by established tests.

Part C of Title V recognizes the need for State Energy Conservation Plans. Congress found that:

1. The development and implementation by states of laws, policies, programs, and procedures to conserve and to improve efficiency in the use of energy will have an immediate and substantial effect in reducing the rate of growth of energy demand and in minimizing the adverse social, economic, political, and environmental impacts of increasing energy consumption
2. The development and implementation of energy conservation programs by states will most effectively minimize any adverse economic or employment impact of changing patterns of energy use and meet local economic, climatic, geographic, and other unique conditions and requirements of each state

3. The federal government has a responsibility to foster and promote comprehensive energy conservation programs and practices by establishing guidelines for such programs and providing overall coordination, technical assistance, and financial support for specific state initiatives in energy conservation

The act also addressed Industrial Energy Conservation. The act requires that the Federal Energy Administration identify each major energy-conserving industry in the United States and establish a priority ranking of such industries on the basis of respective total annual energy consumption. Within each industry, the administration identified each corporation that consumed at least one trillion British thermal units of energy each year; and, among the corporations identified by the administration, the 50 most energy consumptive corporations in such industry.

Within one year of enactment, the administration was to set an industrial energy efficiency improvement target for each of the ten most energy consumptive industries.

National Energy Conservation Policy Act of 1978 (Public Law 95-619, 92 Statute 3206)

This act amends and adds to the Energy Policy and Conservation Act of 1975. The act is based on the congressional findings that:

1. The United States faces an energy shortage arising from increasing demand for energy, particularly for oil and natural gas, and insufficient domestic supplies of oil and natural gas to satisfy that demand
2. Unless effective measures are promptly taken by the federal government and other users of energy to reduce the rate of growth of demand for energy, the United States will become increasingly dependent on the world oil market, increasingly vulnerable to interruptions of foreign oil supplies, and unable to provide the energy to meet future needs
3. All sectors of our nation's economy must begin immediately to significantly reduce the demand for nonrenewable energy resources, such as oil and natural gas, by implementing and maintaining effective conservation measures for the efficient use of these and other energy sources

The purposes of this act are to:

1. Provide for the regulation of interstate commerce
2. Reduce the growth in demand for energy in the United States
3. Conserve nonrenewable energy resources produced in this nation and elsewhere, without inhibiting beneficial economic growth

A number of titles provide the means to implement this act. Title II is devoted to Residential Energy Conservation. Energy conservation measures are defined. Tax incentives are stipulated in the Energy Tax Act of 1978. Specific aspects of Title II include Weatherization Grants for the Benefit of Low-Income Families, Secondary Financing and Loan Insurance for Energy-Conserving Improvements and Solar Energy Systems, Energy-Conserving Improvements for Assisted Housing, and Energy-Conserving Standards for Newly Constructed Residential Housing Insured by the Federal Housing Administration.

Title III consisted of Energy Conservation Programs for Schools and Hospitals and Buildings Owned by Units of Local Governments and Public Care Institutions. Congress found that:

1. Schools and hospitals are major consumers of energy, and have been especially burdened by rising energy prices and fuel shortages
2. Substantial energy conservation can be achieved in schools and hospitals through the implementation of energy conservation, maintenance, and operating procedures, and the installation of energy conservation measures
3. Public and nonprofit schools and hospitals in many instances need financial assistance in order to make the necessary improvements to achieve energy conservation

The purpose of Title III is to:

1. Authorize grants to states and to public and nonprofit schools and hospitals to assist them in identifying oil implementing energy conservation maintenance and operating procedures

2. Evaluate energy conservation measures
3. Reduce the energy use and anticipated energy costs of
    schools and hospitals

To implement this act, guidelines were prepared by the secretary of energy. The states were then requested to submit energy audits and list the types of energy conservation projects considered appropriate for schools and hospitals. Grants for project costs and technical assistance are available.

Part 2 of Title III offered the same conservational measures to units of local governments and public institutions.

Title IV is devoted to Energy Efficiency of Certain Products and Processes. Part I considered Energy Efficiency Standards for Automobiles; Part II, fuel standards for Consumer Products other than automobiles; and, Part III considered Industrial Equipment. This title was based on the concept that:

1. The average per household energy use within the
    United States by product exceeded 150 kilowatt-hours
    for any 12-month period
2. The aggregate household energy use within the United
    States by product exceeded 4.2 billion kilowatt-hours for
    any 12-month period
3. Substantial improvement in the energy efficiency of
    products is technologically feasible

Title V considered Federal Energy Initiatives for a number of energy sources. In order to encourage solar energy utilization, the act established procedures for the development of demonstration solar heating and cooling in federal buildings; and, in addition, the implementation of energy conservation and solar energy in federal buildings was based on congressional findings that:

1. There is an urgent need to promote the design,
    construction, and operation of buildings to conserve
    and make more efficient use of fuels and energy
2. A shift from dependence on nonrenewable to renewable
    energy sources would have a beneficial effect on the
    nation's overall energy supply
3. Programs for energy conservation in buildings, along
    with the use of renewable energy sources, could
    stimulate industries and create new job opportunities

    for supply and servicing new or improved energy-conserving and energy-supplying systems and equipment

4. In the construction or renovation of buildings, the cost of energy consumed over the life of such buildings must be considered as well as the initial cost of such construction or renovation

5. The federal government, the largest energy consumer in the United States, should be in the forefront in implementing energy conservation measures and in promoting the use of solar heating and cooling and other renewable energy sources

From these findings the stated policy of the United States is that the federal government has the opportunity and responsibility, with the participation of industry, to further develop, demonstrate, and promote the use of energy conservation, solar heating and cooling, and other renewable energy sources in federal buildings.

Part IV of Title V is cited as the Federal Photovoltaic Utilization Act. This act established a photovoltaic energy commercialization program for the accelerated procurement and installation of photovoltaic solar electric systems for electric production in federal facilities.

Title VI considered a number of conservational measures. These included industrial energy efficiency reporting, state energy conservation plans, minority economic impact, and conservation of national coal resources.

Energy Emergency Preparedness Act of 1982 (Public Law 97-229, 96 Statute 248)

The purpose of this measure is to amend the Energy Policy and Conservation Act, extend certain authorities relating to the Instructional Energy Program, provide for the nation's energy emergency preparedness, and related areas.

Congress found that Energy Emergency Preparedness was required because:

1. A shortage of petroleum products caused by reductions in imports may occur at anytime

2. Such a shortage may be sufficiently large to cause severe economic distortions and hardships, or constitute a serious threat to public health, safety, and welfare
3. Prior to the occurrence of such a shortage, the federal government has a responsibility to be prepared to mitigate the adverse impacts of such a shortage as a supplement to reliance on free market pricing and allocation of available petroleum product supplies

This act establishes the policy that the federal government will be prepared, prior to any shortage of petroleum products, to respond to energy emergencies. To implement this program, the president was required to submit to Congress comprehensive energy response procedures. This plan shall:

1. Describe the various options the president would consider in responding to a severe energy supply interruption
2. Specify how appropriate governmental actions in response to international and domestic energy shortages would be selected and implemented
3. Recommend any additional statutory authority the president considers necessary

The Strategic Petroleum Reserve Amendments specify that:

After the Strategic Petroleum Reserve reaches a level of 5 million barrels, the president shall immediately seek to undertake, and thereafter continue, petroleum products acquisition, transportation, and injection activities at a level sufficient to assure that the petroleum products in the Strategic Petroleum Reserve will increase at an average arrival rate of at least 300,000 barrels per day until the quantity of petroleum products stored within the Strategic Petroleum Reserve is at least 750 million barrels.

The act also required the secretary of the Department of Energy to analyze the impact on the domestic economy and on United States' consumers of reliance on market allocation and pricing during any substantial reduction in the amount of petroleum products available in the United States.

International Appliance Energy Conservation Act of 1987 (Public Law 100-12, 101 Statute 103)

This act established standards and tests to measure the efficiency of appliances. Appliances, such as refrigerators, freezers, air conditioners, water heaters, pool heaters, direct heating equipment, furnaces, clothes driers, kitchen ranges and ovens, and many others are included.

The term "energy conservation" in the act means:

1. A performance standard which prescribes a minimum level of energy efficiency on a maximum quantity of energy use for a covered product, determined by an established test procedure
2. Development of a design requirement for the products specified

The standards are to be evaluated no later than January 1, 1992, to determine if the initial standards established in the act need to be amended.

Public Utilities Regulatory Policies Acts

November 9, 1978, Public Law 95-617, 92 Statute 3117

June 30, 1980, Public Law 96-294, 94 Statute 718, 770

November 8, 1984, Public Law 98-620, 98 Statute 3360

October 16, 1986, Public Law 99-495, 100 Statute 1249

November 15, 1990, Public Law 101-575, 104 Statute 2834

Public Utilities Regulatory Policies Act of 1978 (Public Law 95-617, 92 Statute 3117)

Congress passed the act in order to regulate interstate commerce in energy requirements for the protection of public health, safety, and welfare, and the preservation of natural security. The act provided for a program:

1. For increased conservation of electric energy, increased efficiency in the use of facilities and resources of electric utilities and equitable retail rates for electric consumers

2. To improve the wholesale distribution of electric energy, the reliability of electric service, and procedures for wholesale rate applications before the Federal Energy Regulatory Commission
3. To provide for the expeditious development of hydroelectric potential at existing small dams
4. For the conservation of natural gas while ensuring that rates to natural gas consumers are equitable
5. To encourage the development of crude oil transportation systems

Energy Efficiency Legislation

Federal Energy Administration Acts

May 7, 1974, Public Law 93-275, 88 Statute 96

June 30, 1976, Public Law 94-332, 90 Statute 784

August 14, 1976 , Public Law 94-385, 90 Statute 1127-1132, 1135

July 21, 1977, Public Law 95-70, 91 Statute 275, 277-280

August 4, 1977, Public Law 95-91, 91 Statute 607, 608

October 19, 1980, Public Law 96-470, 94 Statute 2244

Federal Energy Administration Act Amendment of 1976 (Public Law 94-385, 90 Statute 1127)

Federal Energy Administration Authorization Act of 1977 (Public Law 95-70, 91 Statute 275)

Federal Energy Management Improvement Act of 1988 (Public Law 100-615, 102 Statute 3185)

This act is to promote the conservation and efficient use of energy by the federal government. In passing the act Congress found that:

1. The federal government is the largest single energy consumer in the nation

2. The cost of meeting the federal government's energy requirement is substantial
3. There are significant opportunities in the federal government to conserve and make more efficient use of energy through improved operations and maintenance, and use of new energy efficient technologies, and the application and achievement of energy efficient design and construction
4. Federal energy conservation measures can be financed at little or no cost to the federal government by using private investment capital made available through contacts authorized by Title VIII of this act
5. An increase in energy efficiency by the federal government would benefit the nation by reducing the cost of government, reducing national dependence on foreign energy resources, and demonstrating the benefits of greater energy efficiency to the nation

To implement the act, each agency of the government shall apply energy conservation measures in its building program so that by 1995 energy consumption per gross square feet will be reduced by at least 10 percent. The implementation steps are:

1. Each agency prepares a plan as to how it will meet this goal
2. Perform energy surveys of all federal buildings
3. Enforce energy conservation measures in order to attain the act's goal

In order to develop life cycle cost methods to determine conservation procedures, the General Services Administration shall establish practical and effective present value methods for estimating and comparing life cycle costs for federal buildings.

In order to secure maximum conservation of energy, each agency shall establish a program of incentives for conserving energy. An Interagency Energy Management Task Force has been established to:

1. Access the progress of the various agencies in achieving energy savings
2. Collect and disseminate information
3. Coordinate energy savings

4. Develop options for use in conserving energy
5. Review regulations relating to building temperature settings to determine if changes are appropriate

The secretary of the Department of Energy is empowered to:

1. Determine the maximum potential cost-effective energy savings that may be achieved in a representation sample of buildings owned or leased by the federal government in different areas of the country
2. Make recommendations for cost-effective energy efficiency and renewable energy improvements

Federal Energy Administration Act of 1974
(Public Law 93-275, 88 Statute 96)

This act reorganized and consolidated certain functions of the federal government into a new Federal Energy Administration in order to promote more effective management.

Congress found that the general welfare and common defense and security of the nation required positive and effective action to conserve source energy supplies; ensure fair and efficient distribution of, and the maintenance of fair and reasonable consumer prices for, energy to promote the expansion of readily usable energy sources; and, assist in developing policies and plans to meet the energy needs of the nation. In order to achieve these objectives and to ensure a coordinated and effective approach to overcoming energy shortages, it is necessary to reorganize certain agencies and functions of the Executive Branch and to establish a Federal Energy Administration.

The functions of the Federal Energy Administration are:

1. To advise the president and the Congress with respect to the establishment of a comprehensive national energy policy
2. To assess the adequacy of energy resources to meet demands in the immediate and longer range future for all sectors of the economy and for the general public
3. To develop effective arrangements for the participation of state and local governments in the resolution of energy problems
4. To develop plans and programs for dealing with energy problems

5. To promote stability in energy prices to the consumer, promote free and open competition in all aspects of the energy field, prevent unreasonable profits within the various segments of the energy industry, and promote free enterprise
6. To ensure that energy programs are designed and implemented in a fair and efficient manner so as to minimize hardship and inequity while ensuring that the priority needs of the nation are met
7. To develop and oversee the implementation of equitable voluntary and mandatory energy conservation programs and promote efficiency in the use of energy resources
8. To develop and recommend policies of the import and export of resources
9. To collect, evaluate, assemble, and analyze energy information on reserves, production, demand, and related economic data
10. To work with business, labor, consumers, and other interests to obtain their cooperation
11. In administering any pricing authority, to develop necessary rules and regulations

Energy Reorganization Acts

October 11, 1974, Public Law 93-438, 88 Statute 1233

August 9, 1975, Public Law 94-79, 89 Statute 413, 414

June 3, 1977, Public Law 95-39, 91 Statute 200

December 13, 1977, Public Law 95-209, 91 Statute 1482

February 25, 1978, Public Law 95-238, 92 Statute 56

November 6, 1978, Public Law 95-601, 92 Statute 2949–2951

August 22, 1986, Public Law 99-386, 100 Statute 822

Energy Reorganization Act of 1974
(Public Law 93-438, 88 Statute 1233)

This act reorganized and consolidated certain functions of the federal government in a new Energy Research and Development Administration and in a new Nuclear Regulatory Commission in order to promote more efficient management.

Energy Research and Development Administration   The responsibilities of the administration include:

1. Exercising central responsibility for policy planning, coordination, support, and management of research and development programs respecting all energy sources, including assessing the requirements for research and development in regard to various energy sources in relation to near-term and long-term needs, policy planning in regard to meeting those requirements, undertaking programs for the optimal development of the various forms of energy sources, managing such programs, and disseminating information resulting therefrom
2. Encouraging and conducting research and development including demonstrations of commercial feasibility and practical applications of the extraction, conversion, storage, transmission, and utilization phases related to the development and use of energy from fossil, nuclear, solar, geothermal, and other sources
3. Engaging in and supporting environmental, biomedical, physical and safety research related to the development of energy sources and utilization technologies
4. Taking into account the existence, progress, and results of other public and private research and development activities relevant to the administration's mission in formulating its own research and development program
5. Participating in and supporting cooperative research and development projects
6. Developing, collecting, distributing, and making available for distribution, scientific and technical information concerning the manufacture or development of energy and its efficient extraction, conversion, transmission, and utilization
7. Creating and encouraging the development of general information to the public on all energy conservation technologies and energy sources as they become available for use
8. Encouraging and conducting research and development in energy conservation
9. Encouraging and participating in international cooperation in energy and related environmental research and development

10. Helping to ensure an adequate supply of manpower for the accomplishment of energy research and development programs
11. Encouraging and conducting research and development in clean and renewable energy sources

Nuclear Regulatory Commission The Atomic Energy Commission established under the Atomic Energy Act of 1954 was abolished and all of its functions transferred to the new Nuclear Regulatory Commission.

Each year the commission reports to the president on its activities and findings in the following areas:

1. Ensuring the safe design of nuclear power plants and other licensed facilities
2. Investigating abnormal occurrences and defects in nuclear power plants and other licensed facilities
3. Safeguarding special nuclear materials at all stages of the nuclear fuel cycle
4. Investigating suspected, attempted, or actual thefts of special nuclear materials in the licensed sector and developing contingency plans for dealing with such incidences
5. Ensuring the safe, permanent disposal of high-level radioactive wastes through the licensing of nuclear activities and facilities
6. Protecting the public against the hazards of low-level radioactive emissions from licensed nuclear activities and facilities

A major feature of the act was the establishment of a Nuclear Energy Center Site Survey. The commission was authorized to conduct a national survey, which included consideration of each of the existing or future electric reliability regions, to locate and identify possible nuclear energy center sites.

The survey shall include:

1. A regional evolution of natural resources, including land, air, and water resources, available for use in connection with nuclear energy center sites, estimates of future electric power requirements that can be secured

by each nuclear energy center site, an assessment of the economic impact of each nuclear energy site, and consideration of other factors
2. An evaluation of the environmental impact likely to result from construction and operation of such nuclear energy centers, including an evaluation as to whether such nuclear energy centers will result in greater or lesser environmental impact than separate siting of the reactors and/or fuel cycle facilities
3. Consideration of the use of federally owned property and other property designated for public use, but excluding national parks, national forests, national wilderness areas, and national historic monuments

Energy Conservation and Production Acts

August 14, 1976, Public Law 94-385, 90 Statute 1125

July 21, 1977, Public Law 95-70, 91 Statute 277

November 6, 1978, Public Law 95-602, 92 Statute 2987

November 9, 1978, Public Law 95-617, 92 Statute 3133

October 19, 1980, Public Law 96-470, 94 Statute 2243

December 21, 1982, Public Law 97-375, 96 Statute 1824

Energy Conservation Standards for New Buildings Acts

August 14, 1976, Public Law 94-385, 90 Statute 1144–1149

August 4. 1977, Public Law 95-91, 91 Statute 608

November 9, 1978, Public Law 95-619, 92 Statute 3238

October 8, 1980, Public Law 96-399, 94 Statute 1649

August 13, 1981, Public Law 97-35, 95 Statute 621

February 5, 1988, Public Law 100-242, 101 Statute 1950

Energy Conservation Standards for Existing Buildings Acts

August 14, 1976, Public Law 94-385, 90 Statute 1150–1169

November 9, 1978, Public Law 95-619, 92 Statute 3224

June 30, 1980, Public Law 96-294, 94 Statute 759, 760

November 30, 1983, Public Law 98-181, 97 Statute 1235

October 17, 1984, Public Law 98-479, 98 Statute 2228

October 30, 1984, Public Law 98-558, 98 Statute 2887, 2888

February 5, 1988, Public Law 100-242, 101 Statute 1950

Energy Conservation and Production Act of 1976 (Public Law 94-385, 90 Statute 1125)

After study, Congress found that the fastest, most cost-effective, and most environmentally sound way to prevent future energy shortages in the United States, while reducing the nation's dependence on imported energy supplies, is to encourage and facilitate, through major programs, the implementation of energy conservation and renewable resource energy measures with respect to dwellings, nonresidential buildings, and industrial plants.

This act amends the Federal Energy Administration Act of 1974 to extend the duration of authorities from 1976 to 1979, create an incentive for domestic production, provide for electric utility rate design initiatives, provide for energy conservation standards for new buildings, provide for energy conservation assistance for existing buildings and industrial plants, and for other purposes.

This act incorporates a number of other acts under different titles.

Title I, the Federal Energy Administration Act Amendments of 1976, sets policies and procedures. Besides basic administration amendments, Part B considers production enhancement of petroleum wells. Production for stripper wells (wells that produce less than ten barrels of oil per day) is encouraged. Part C establishes a National Energy Information System to ensure the availability of adequate comparable, accurate, and credible energy information to the Federal Energy Administration, to other government agencies responsible for energy-activated policy decisions, to the Congress, and to the public.

Table II treats Electric Utility Rote Design Initiatives. Congress found that improvement in electric utility rate design had great potential for reducing the cost of electric utility services to consumers, and for encouraging energy conservation and better use of existing electrical generating facilities. It was the purpose of this title to require the Federal Energy Administration to develop proposals for improvement of electric utility rate design and transmit such proposals to Congress, to fund electric utility rate demonstration projects, to intervene or participate upon request in the proceedings of utility regulatory commissions, and to provide janitorial assistance to state offices of consumer services to facilitate presentation of consumer interests before such commissions.

Title III, cited as the Energy Conservation Standards for New Buildings Act of 1976, was based on findings that:

1. Large amounts of fuel and energy are consumed unnecessarily each year in heating, cooling, ventilating, and providing domestic hot water for newly constructed residential and commercial buildings because such buildings lack adequate energy conservation features
2. Federal performance standards for newly constructed buildings can prevent such energy waste, which the nation can no longer afford in view of its current and anticipated energy shortages
3. The failure to provide adequate energy conservation measures in newly constructed buildings increases long-term operating costs that may affect adversely the repayment of, and security for, loans made, insured, or guaranteed by federal agencies
4. State and local building codes or similar controls can provide an existing means by which to ensure, in coordination with other building requirements and with a minimum of federal interference on state transactions, that newly constructed buildings contain adequate energy conservation features

The purpose of this title was to:

1. Redirect federal policies and practices to ensure that reasonable energy conservation features will be incorporated into new commercial and residential buildings receiving federal financial assistance
2. Provide for the development and implementation, as soon as practicable, of performance standards for new residential and commercial buildings, which are

> designed to achieve the maximum practicable improvement in energy efficiency and increases in the use of nondepletable sources of energy
>
> 3. Encourage state and local governments to adopt and enforce such standards through their existing building codes and other construction control mechanisms, or apply them through a special approval process

Title IV is cited as the Energy Conservation in Existing Buildings Act of 1976. It is the finding of Congress that:

> 1. Dwellings owned or occupied by low-income persons frequently are inadequately insulated
> 2. Low-income persons, particularly elderly and handicapped, can least afford to make the modifications necessary to provide for adequate insulation in such dwellings and to otherwise reduce residential energy use
> 3. Weatherization of such dwellings would lower utility expenses for such low-income owners or occupants as well as save thousands of barrels per day of needed fuel
> 4. States, through community action agencies established under the Economic Opportunity Act of 1964 and units of general purpose local government, should be encouraged, with federal financial and technical assistance, to develop and support coordinated weatherization programs designed to ameliorate the adverse effects of the high energy costs of such low-income persons, to supplement other federal programs serving such persons, and to conserve energy

The purpose of this act was to develop and implement a weatherization assistance program to assist in achieving a prescribed level of insulation in the dwellings of low-income persons, particularly elderly and handicapped, in order to aid those persons least able to afford higher utility costs and to conserve needed energy.

Energy Tax Acts

November 9, 1978, Public Law 95-618, 92 Statute 3174

April 2, 1980 , Public Law 96-223, 94 Statute 273

September 3, 1982, Public Law 97-248, 96 Statute 575

Energy Tax Act of 1978
(Public Law 95-618, 92 Statute 3174)

This act was to provide tax incentives for the production and conservation of energy and for other purposes.

Title I, Residential Energy Credit   For any dwelling unit, the qualified renewable energy expenditures were the following percentage of the renewable energy source expenditure made by the taxpayer during the taxable year with respect to such units:

1. Thirty percent of such expenditures that does not exceed $2,000, plus
2. Twenty percent of such expenditures that exceeds $2,000 but does not exceed $10,000

Energy Conservation Expenditures include insulation of the house and such other energy-saving devices as:

1. A furnace replacement burner so designed to achieve a reduction in the amount of fuel consumed
2. A device for modifying flue openings designed to increase the efficiency of operations
3. An electrical or mechanical furnace ignition system, which replaces a gas pilot light
4. A storm or thermal window or door for the exterior of the dwelling
5. An automatic energy-saving set-back thermostat
6. Caulking or weatherstripping of an exterior door or window
7. A meter that displays the cost of energy usage

Title II, Transportation   The basic feature of Title II was the imposition of the Gas-Guzzler Tax. In the case of a 1980 model year automobile, taxes were determined by the fuel economy of the model type in which the automobile falls, as follows:

Fuel Economy	Tax
At least 15 miles per gallon	$0
At least 14 but less than 15	$200
At least 13 but less than 14	$300
Less than 13	$500

By 1986 the fuel economy and tax was:

Fuel Economy	**Tax**
At least 22.5 miles per gallon	$0
At least 21.5 but less than 22.5	$500
At least 18.5 but less than 19.5	$1,050
At least 15.5 but less than 16.5	$1,850
At least 12.5 but less than 13.5	$3,200
Less than 12.5	$3,850

The fuel economy for any model type was increased with testing and calculation procedures established by the Environmental Protection Agency.

A provision was also made for exemption from motor fuels excise tax and for certain alcohol fuels. No excise tax was to be imposed on the sale of any gasoline:

1. In a mixture with alcohol, if at least 10 percent of the mixture was alcohol
2. For use in producing a mixture at least 10 percent of which is alcohol

Incentives were also provided for car pooling. A commuter highway vehicle means one that has:

1. Been acquired by the taxpayer after the enactment of the Energy Tax Act of 1978, and placed in service by the taxpayer before January 1, 1986
2. Seating capacity, which is at least eight adults (not including the driver)
3. At least 80 percent of the mileage use could reasonably be expected to be for purposes of transporting employees between their residences and their place of employment, and on trips during which the number of employees transported for such purposes is at least one-half of the adult seating capacity of such vehicle (not including the driver)

Title III considered Changes in Business Investment Credit to Encourage Conservation of, or Conversion from, Oil and Gas and to Encourage New Energy Technology.

Crude Oil Windfall Profit Tax Acts

April 2, 1980, Public Law 96-223, 94 Statute 229

October 19, 1980, Public Law 96-471, 94 Statute 2253–2254

August 13, 1981, Public Law 97-34, 95 Statute 272

October 25, 1982, Public Law 97-362, 96 Statute 1726

January 12, 1983, Public Law 97-448, 96 Statute 2396

Crude Oil Windfall Profit Tax of 1980
(Public Law 96-223, 94 Statute 229)

This act was passed because of the administration's decision to phase out price controls on crude oil at a time when oil prices were rising and the nation continued its over dependence on imported oil. This legislation provided the means to tax a fair share of the additional revenues received by oil producers and royalty owners, and at the same time not to reduce incentives to increase domestic production.

Title II of the act provided "Energy Conservation and Production Incentives." These included residential energy credit by providing specifications to reduce oil and gas consumption, business energy investment credit including development of renewable energy, the production of fuel from nonconventional sources such as alcohol fuels, and energy-related uses of tax-exempt bonds such as investments in solid waste disposal facilities, hydroelectric generating facilities and renewable energy property.

Title III develops policies for low-income energy assistance, such as eligible households, state plans, allotments, home energy grants and payments.

Atomic Energy Legislation

Atomic Energy Act of 1946

August 1, 1946, Chapter 724, 60 Statute 755

July 3, 1948, Chapter 828, 62 Statute 1259

October 11, 1949, Chapter 673, 63 Statute 762

September 23, 1950, Chapter 1000, 64 Statute 979

October 30, 1951, Chapter 633, 65 Statute 692

April 5, 1952, Chapter 159, 66 Statute 43

July 7, 1953, Chapter 228, 67 Statute 181

July 31, 1953, Chapter 283, 67 Statute 240

July 31, 1953 , Chapter 284, 67 Statute 240

August 13, 1953, Chapter 432, 67 Statute 575

August 13, 1954, Chapter 730, 68 Statute 715

August 30, 1954 , Chapter 1073, 68 Statute 919

Atomic Energy Act of 1946
(Chapter 724, 60 Statute 755)

This act provided the basic framework for the development of atomic energy. In Section 1, "Findings and Declarations," it states:

> Research and experimentation in the field of nuclear chain reaction have attained the stage at which the release of atomic energy on a large scale is practical. The significance of the atomic bomb for military purposes upon the social, economic, and political structures of today cannot now be determined. It is a field in which unknown factors are involved. Therefore any legislation will necessarily be subject to revision from time to time. It is reasonable to anticipate, however, that tapping this new source of energy will cause profound changes in our present way of life. Accordingly, it is hereby declared to be the policy of the people of the United States that, subject at all times to the paramount objectives of assuring the common defense and security, the development and utilization of atomic energy shall, so far as practicable, be directed toward improving the public welfare, increasing the standard of living, strengthening free competition in private enterprise, and promoting world peace.

The purpose of the act to implement the above policies included the following:

1. A program of assisting and fostering private records and development to encourage maximum scientific progress
2. A program for the control of scientific and technical information that will permit the dissemination of such information to encourage scientific progress, and for

sharing information on a reciprocal basis concerning the practical industrial application of atomic energy as soon as effective and enforceable safeguards against its use for distribution purposes can be devised
3. A program of federally conducted research and development to assure the government of adequate scientific and technical accomplishment
4. A program for government control of the production, ownership, and use of fissionable material to ensure the common defense and security and to ensure the broadest possible exploitation of the fields
5. A program of administration that will be consistent with the foregoing policies and with international arrangements made by the United States, and which will enable Congress to be currently informed so as to take further legislative action as may be appropriate

The 1946 act established the Atomic Energy Commission comprised of five members. In addition there was a General Advisory Committee and a Military Liaison Committee. The General Advisory Committee gave advice to the commission on scientific and technical matters relating to materials, production, research, and development. The Military Liaison Committee kept the commission informed of all atomic energy activities of the defense establishments.

The commission had responsibility for:

Research   The commission is directed to ensure the continued conduct of research and development activities by private or public institutions or persons, and to assist in the acquisition of an ever-expanding amount of theoretical and practical knowledge in such fields. This includes activities relating to:

1. Nuclear processes
2. Theory and production of atomic energy
3. Utilization of fissionable and radioactive material for medical, biological, health, or military purposes
4. Utilization of fissionable and radioactive materials and processes entailed in the production of such materials for all these purposes, including industrial uses
5. The protection of health during research and production activities

Production of Fissionable Material   The commission is the exclusive owner of all facilities for the production of fissionable material other than facilities which: (1) are useful in the conduct of research and development activities, and (2) do not, in the opinion of the commission, have a potential productive rate adequate to produce an atomic bomb.

Control of Materials   The U.S. government is the sole owner of all fissionable material. The commission controls the distribution of all atomic materials and, by this act, established standards for the issuance, refusal, or revocation of licenses as it may deem necessary in order to ensure adequate source material.

Military Application of Atomic Energy   The commission is authorized to conduct experiments and do research and development work in the military applications of atomic energy and engage in the production of atomic bombs, atomic bomb parts, and other military weapons.

Utilization of Atomic Energy   The act states that whenever in its opinion any industrial, commercial, or other nonmilitary use of fissionable material or atomic energy has been sufficiently developed to be of practical value, the commission shall report to the president stating all the facts with respect to such use; the commission's estimate of the social, political, economic, and international effects of such use; and the commission's recommendations for necessary or desirable supplementary legislation.

International Arrangements   "International Arrangements" meant any treaty approved by the Senate or international agreement approved by the Congress, during the time such treaty agreement is in effect. Any provision of this act that conflicts with the provision of any international arrangement made after the date of enactment of this act shall be deemed to be of no further force or effect.

Control of Information   It is the policy of the commission to control the dissemination of restricted data in such a manner as to ensure the common defense and security. The following principles guide the commission:

1. Until enforceable international safeguards are established, there is to be no exchange of information with other nations with respect to the use of atomic energy for industrial purposes

2. The dissemination of scientific and technical information relating to atomic energy shall be permitted and encouraged to provide free interchange of ideas and criticisms essential to scientific progress

Patents and Inventions   No patents are to be granted for any invention or discovery which was useful solely in the production of fissionable material or in the utilization of fissionable material or atomic energy for a military weapon.

General Authority   The commission is authorized to:

1. Establish advisory boards to consult with and make recommendations on legislation, policies, administration, research, and other matters
2. Make studies and investigations, obtain such information, and hold hearings to assist in administering the act
3. Establish by regulation or order such standards and instruction to govern the possession and use of fissionable and by-product materials
4. Acquire such material, property, equipment and facilities, and establish or construct such buildings and facilities as needed
5. Use the personnel of any government agency

Enforcement   The commission has the responsibility to enforce the provisions of this act by permanent or temporary injunction, restraining order, or any other court order.

Atomic Energy Act of 1954

August 30, 1954, Chapter 1073, 68 Statute 919

August 9, 1955, Chapter 697, 69 Statute 630

July 14, 1956, Chapter 608, 70 Statute 553

August 6, 1956, Chapter 1015, 70 Statute 1069

April 12, 1957, Public Law 85-14, 71 Statute 11

July 3, 1957, Public Law 85-79, 71 Statute 274

August 21, 1957, Public Law 85-162, 71 Statute 410

August 28, 1957, Public Law 85-177, 71 Statute 455

September 2, 1957, Public Law 85-256, 71 Statute 576

September 4, 1957, Public Law 85-287, 71 Statute 612

July 2, 1958, Public Law 85-479, 72 Statute 276

July 7, 1958, Public Law 85-507, 72 Statute 337

August 8, 1958, Public Law 85-602, 72 Statute 525

August 19, 1958, Public Law 85-681, 72 Statute 632

August 23, 1958, Public Law 85-744, 72 Statute 837

June 11, 1959, Public Law 86-43, 73 Statute 73

June 23, 1959, Public Law 86-50, 73 Statute 87

September 21, 1959, Public Law 86-300, 73 Statute 574

September 23, 1959, Public Law 86-373, 73 Statute 688

September 6, 1961, Public Law 87-206, 75 Statute 476

August 29, 1962, Public Law 87-615, 76 Statute 409

October 11, 1962, Public Law 87-793, 76 Statute 864

July 22, 1963, Public Law 88-72, 77 Statute 88

March 26, 1964, Public Law 88-294, 78 Statute 172

August 1, 1964, Public Law 88-394, 78 Statute 376

August 14, 1964, Public Law 88-426, 78 Statute 423,
429, 430

August 19, 1964, Public Law 88-448, 78 Statute 490

August 26, 1964, Public Law 88-489, 78 Statute 602–607

August 24, 1965, Public Law 89-135, 79 Statute 551

September 29, 1965, Public Law 89-210, 79 Statute 855

October 13, 1966, Public Law 89-645, 80 Statute 891–893

December 14, 1967, Public Law 90-190, 81 Statute
577, 578

December 24, 1969, Public Law 91-161, 83 Statute
444, 445

October 15, 1970, Public Law 91-452, 84 Statute 930

December 19, 1970, Public Law 91-560, 84 Statute 1472–1474

August 11, 1971, Public Law 92-84, 85 Statute 307

June 2, 1972, Public Law 92-307, 86 Statute 191

June 16, 1972, Public Law 92-314, 86 Statute 227

May 10, 1974, Public Law 93-276, 88 Statute 119

August 17, 1974, Public Law 93-377, 88 Statute 473–475

October 26, 1974, Public Law 93-485, 88 Statute 1460

December 6, 1974, Public Law 93-514, 88 Statute 1611

December 27, 1974, Public Law 93-554, 88 Statute 1776

December 31, 1975, Public Law 94-197, 89 Statute 1111

September 20, 1977, Public Law 95-110, 91 Statute 884

December 3, 1977, Public Law 95-209, 91 Statute 1483

March 10, 1978, Public Law 95-242, 92 Statute 125–148

November 6, 1978, Public Law 95-601, 92 Statute 2950

November 9, 1979, Public Law 96-106, 93 Statute 800

June 30, 1980, Public Law 96-295, 94 Statute 786–789

December 12, 1980, Public Law 96-517, 94 Statute 3027

December 4, 1981, Public Law 97-90, 95 Statute 1169, 1170

April 2, 1982, Public Law 97-164, 96 Statute 48, 49

January 4, 1983, Public Law 97-415, 96 Statute 2071, 2073, 2075–2081

November 8, 1984, Public Law 98-622, 98 Statute 3388

July 12, 1985, Public Law 99-64, 99 Statute 159

August 26, 1986, Public Law 99-399, 100 Statute 875, 876

November 14, 1986, Public Law 99-661, 100 Statute 4064

August 20, 1988, Public Law 100-408, 102 Statute 1066–1083

September 28, 1988, Public Law 100-449, 102 Statute 1876

September 29, 1988, Public Law 100-456, 102 Statute 2076

November 29, 1989, Public Law 101-189, 103 Statute 1684

November 5, 1990, Public Law 101-510, 104 Statute 1844

November 15, 1990, Public Law 101-575, 104 Statute 2835

November 29, 1990, Public Law 101-647, 104 Statute 4833

December 5, 1991, Public Law 102-190, 105 Statute 1582

Atomic Energy Act of 1954
(Public Law 703, Chapter 1073, 68 Statute 919)

The primary purpose of the 1954 act was to bring the atomic energy list of 1946 into accord with atomic programs and to make Congress's legislative controls conform with the scientific, technical, economic, and political facts of atomic energy in 1954.

The organic law was written at the very outset of the atomic era. The legislators were keenly aware that there were many unknown factors involved in measuring the impact of this new source of energy. Under the 1946 act, while stress was placed on developing the atomic weapon stockpile, there were developments beneficial to society. Still 95 percent of atomic energy was used for military purposes and possibly 5 percent for peacetime uses. In 1946 it was believed that the generation of useful power from atomic energy was a distant goal. By 1954, however, with the experience acquired in designing, building, and operating more than a score of atomic reactors, it became evident that power generation is possible and need not bring attendant hazards to the health and safety of the American people. While many technical problems remained to be solved in 1954, it was evident that with the help of private enterprise atomic energy could be achieved quickly.

The 1954 act set the stage for the peaceful development of atomic power. It declared that the policy of the United States was:

1. The development, use, and control of atomic energy shall be directed so as to make the maximum contribution to the general welfare, subject at all times to the paramount objective of making the maximum contribution to the common defense and security

2. The development, use, and control of atomic energy
   shall be directed so as to promote world peace, improve
   the general welfare, increase the standard of living, and
   strengthen free competition in private enterprise

With the passage of the 1954 act, Congress listed the following findings of concern to the development, use, and control of atomic energy:

1. The development, utilization, and control of atomic
   energy for military and for all other purposes vital to
   the common defense and security
2. In permitting the property of the United States to be
   used by others, such use must be regulated in the
   national interest
3. The processing and utilization of source, by-product,
   and special nuclear material effect interstate and
   foreign commerce and must be regulated in the
   national interest
4. The processing and utilization of source, by-product,
   and special nuclear material must be regulated in the
   national interest
5. Source and special nuclear material, production
   facilities, and utilization facilities are affected by the
   public interest, and control by the United States is
   necessary in the national interest to ensure the common
   defense and security and to protect the health and
   safety of the public
6. The necessity for protection against possible interstate
   damage occurring from the operation of facilities for
   the production or utilization of sources or special
   nuclear material places the operation of those facilities
   in interstate commerce for the purpose of this act
7. Funds may be provided for the development and use of
   atomic energy
8. It is essential to the common defense and security that
   title to all special nuclear material be controlled by the
   United States government while such special nuclear
   material is within the United States

The purpose of the act was to implement the above policies by providing a program:

1. Of conducting, assisting, and fostering research and development in order to encourage maximum scientific and industrial progress
2. For the dissemination of unclassified scientific and technical information and for the control, dissemination, and declassification of restricted data
3. For government control of the possession, use, and production of atomic energy and special nuclear material
4. To encourage widespread participation in the development and utilization of atomic energy for peaceful purposes to the maximum extent consistent with the common defense and security and with the health and safety of the public
5. A program of international cooperation to promote the common defense and security
6. A program of administration which will be consistent with the foregoing policies and programs

Since atomic energy technology has advanced and public opinion has developed, the 1954 act has been amended many times. A listing of these amendments is given for those who desire to study the evolution of this act. To illustrate, in 1989 under the Defense Authorization Act, Title XXXII, Defense Nuclear Facilities Safety Board Authorization, it states:

> These are authorized to be appropriated for fiscal year 1990, $7 million for the establishment and operation of the Defense Nuclear Facilities Safety Board under Chapter 21 of the Atomic Energy Act of 1954.

Atomic Energy Damage Act of 1957
(Public Law 85-256, 71 Statute 576)

When the Atomic Energy Act of 1954 was passed it was the hope of Congress that the government monopoly in the atomic energy field would encourage the entrance of private industry into the program. By 1956 Congress felt that the danger in the operation of reactors had become a major deterrent to further industrial participation.

As a consequence the Atomic Energy Act of 1954 was amended as follows:

In order to protect the public and to encourage the development of the atomic energy industry, in the interest of the general welfare and of the common defense and security, the United States may make funds available for a portion of the damages suffered by the public from nuclear incidents, and may limit the liability to those persons liable for such losses.

It further stated:

The licensee will hold the United States and the commission [Atomic Energy] harmless from any damages resulting from the use or possession of special nuclear materials by the licensee.

The aggregate liability for a single nuclear incident of persons indemnified, including the reasonable costs of investigating and settling claims and defending suits for damage, shall not exceed the sum of $500 million together with the amount of financial protection required of the licensee or contractor.

Atomic Energy Community Acts

August 4, 1955, Chapter 543, 69 Statute 471

July 25, 1956, Chapter 731, 70 Statute 653

August 21, 1957, Public Law 85-162, 71 Statute 410

August 30, 1961, Public Law 87-174, 75 Statute 409

September 28, 1962, Public Law 87-719, 76 Statute 664

August 1, 1964, Public Law 88-394, 78 Statute 376

May 25, 1967, Public Law 90-19, 81 Statute 23

December 14, 1967, Public Law 90-190, 81 Statute 575

December 31, 1975, Public Law 94-187, 89 Statute 1007

February 25, 1978, Public Law 95-238, 92 Statute 60

November 14, 1986, Public Law 99-661, 100 Statute 4066

Atomic Energy Community Act of 1955
(Chapter 543, 69 Statute 471)

When the atomic energy projects were started by the Manhattan Engineering District at Oak Ridge, Tennessee, and Richland, Washington, they were isolated with no residential facilities.

Therefore, due to security, Congress decided that towns would have to be built, not only for the plants, but for the residents. This policy of the federal government, however, was soon abandoned, and the new atomic energy sites at Aikens, South Carolina, Portsmouth, Ohio, and Paducah, Kentucky, were not built by the Atomic Energy Commission.

This act specifically delegated to each county the responsibility to:

1. Facilitate the establishment of local self-government
2. Provide for the orderly transfer to local entities of municipal functions, installations, and utilities
3. Provide an orderly sale to private purchasers of property within those communities with a minimum of dislocation

The purpose of this act was to provide for:

1. The maintenance of conditions that will not impede the recruitment and retention of personnel essential to the atomic energy program
2. The obligation of the United States to contribute to the support of municipal functions in a manner commensurate with the fiscal problems peculiar to the communities by reason of their constriction as natural defense installations, and the municipal and other burdens imposed by governmental or other entities and the communities by the United States in its operation at or near the communities
3. The opportunity for the residents of the communities to assume the obligations and privileges of local self-government
4. The encouragement of the construction of new homes at the communities

National Safety Research, Development, and Demonstration Act of 1980 (Public Law 96-567, 94 Statute 3329)

This act provided for an accelerated and coordinated program of light water nuclear reactor safety research, development, and demonstration to be carried out by the Department of Energy.

Congress based this act on its findings that:

1. Nuclear energy is one of the two major energy sources available for electric energy production in the United States during the balance of the twentieth century
2. Continued development of nuclear power is dependent upon maintaining an extremely high level of safety in the operation of nuclear plants, and in the public recognition that these facilities do not constitute a significant threat to human health and safety
3. It is the responsibility of utilities, as owners and operators of nuclear power plants, to assume that such plants are designed and operated safely and reliably
4. A proper role of the federal government in ensuring nuclear power safety, in addition to regulatory functions, is the conduct of a research, development, and demonstration program to provide important scientific and technical information which contributes to safe operation of the plant

The objective of a research, development, and demonstration program is to:

1. Reduce the likelihood and severity of potentially serious nuclear power plant accidents
2. Reduce the likelihood of disrupting the population in the vicinity of nuclear power plants as a result of nuclear power plant accidents

It is the responsibility of the secretary of energy to:

1. Further define the assessment of risk factors associated with the generic design and operation of nuclear power plants
2. Develop potentially cost-beneficial changes in the generic design and operation of the nuclear power plant
3. Identify the effect of total or partial automation of generic plant systems on reactor safety
4. Conduct further experimental investigations under abnormal operational and postulated accident conditions in order to determine if the plant should continue
5. Provide for the examination and analysis of any nuclear power plant fuel

6. Identify the aptitude, training, and manning levels which are necessary to ensure reliable operator performance under normal, abnormal, and emergency conditions

Nuclear Waste Policy Acts

January 7, 1983, Public Law 97-425, 96 Statute 2201

December 22, 1987, Public Law 100-203, 101 Statute 1330

October 18, 1988, Public Law 100-507, 102 Statute 2541

Nuclear Waste Policy Amendment

December 22, 1987, Public Law 100-203, 101 Statute 1330

Nuclear Waste Policy Act of 1982
(Public Law 97-425, 96 Statute 2201)

The basic purpose of this act was to provide for the development of repositories for the disposal of high-level radioactive waste and spent nuclear fuel, to establish a program of research, development, and demonstration regarding the disposal of high-level radioactive waste and spent nuclear fuel.

In the hearings for the act Congress found that:

1. Radioactive waste creates potential risks and requires safe and environmentally acceptable methods of disposal
2. A national problem has been created by the accumulation of: (1) spent nuclear fuel from nuclear reactors, and (2) radioactive waste from reprocessing of spent nuclear fuel and activities related to medical research and treatment
3. Federal effort during the past 30 years to devise a permanent solution to the problems of civilian radioactive waste disposal have not been adequate
4. Although the government has responsibility for nuclear waste disposal, the costs should be the responsibility of the generator and owner of such waste
5. The generators and owners of the nuclear waste also must bear the cost of the temporary storage of this material

6. State and public participation in the planning and development of repositories is essential in order to promote public confidence
7. High-level radioactive waste and spent nuclear fuel have become major subjects of public concern, and appropriate measures must be taken to ensure public health and safety and maintain the environment for this and future generations

The purposes of the act are to:

1. Establish a schedule for the siting, construction, and operation of repositories that will provide a reasonable assurance that the public and the environment will be adequately protected from high-level radioactive waste
2. Establish federal responsibility and define federal policy
3. Define the relationships between the federal and state governments
4. Establish a Nuclear Waste Fund comprised of payments made by the generators and owners of high-level radioactive waste

Nuclear Nonproliferation Acts

March 10, 1978, Public Law 95-242, 92 Statute 120

November 14, 1986, Public Law 99-661, 100 Statute 4004

Nuclear Nonproliferation Act of 1978
(Public Law 95-242, 92 Statute 120)

This act is to provide for more efficient and effective control over the proliferation of nuclear explosive capacity.
Congress found that:

The proliferation of nuclear explosive devices, the direct capability to manufacture or acquire such devices poses a grave threat to the security interests of the United States and the national efforts to secure world peace.

As a consequence, the policy of the United States is:

1. Actively pursue, through international initiatives, mechanisms for fuel supply assurances and to establish

more effective international control over the transfer and use of nuclear materials, equipment, and nuclear technology for peaceful purposes
2. Take such actions as are required to confirm the reliability of the United States in meeting the commitments to supply nuclear reactors and fuel to actions which adhere to effective nonproliferation policies by establishing procedures to facilitate the timely processing of requests for subsequent arrangements and export licenses
3. Strongly encourage nations which have not ratified the Treaty on the Nonproliferation of Nuclear Weapons to do so at the earliest possible date
4. Cooperate with foreign nations in identifying and adopting suitable technologies for energy production and, particularly, to identify alternative options to nuclear power in aiding such nations to meet their energy needs

The purpose of the act is to:

1. Establish a more effective framework for international cooperation
2. Authorize the United States to take such actions as are required to ensure that it will act reliably in meeting its commitments to supply nuclear reactors and fuel to nations which adhere to the nonproliferation treaty
3. Provide incentives to other nations of the world to join in such international cooperative efforts and to ratify the treaty
4. Ensure effective control by the United States over its export of nuclear materials and equipment, and of nuclear technology

Natural Gas Legislation

Natural Gas Acts

June 21, 1938, Chapter 556, 52 Statute 821

February 7, 1942, Chapter 49, 56 Statute 83

July 25, 1947, Chapter 333, 61 Statute 459

March 27, 1954, Chapter 115, 68 Statute 36

August 28, 1958, Public Law 85-791, 72 Statute 947

May 21, 1962, Public Law 87-454, 76 Statute 72

October 15, 1970, Public Law 91-452, 84 Statute 929

November 9, 1978, Public Law 95-617, 92 Statute 3173

October 6, 1988, Public Law 100-474, 102 Statute 2302

Natural Gas Act of 1988
(Public Law 100-474, 102 Statute 2302)

The Natural Gas Act requires the Federal Energy Regulatory Commission to regulate natural gas companies that sell gas and provide services in interstate commerce.

Natural Gas Policy Acts

November 9, 1978, Public Law 95-621, 92 Statute 3350

May 21, 1987, Public Law 100-42, 101 Statute 314

September 22, 1988, Public Law 100-439, 102 Statute 1720

July 26, 1989, Public Law 101-60, 103 Statute 157, 158

Natural Gas Policy Act of 1978
(Public Law 95-121, 92 Statute 3350)

The purpose of this act was to establish wellhead pricing of natural gas, the decontrol of certain reduced gas prices, incremental pricing, and emergency pricing.

The act also considers Natural Gas Curtailment Policies for agricultural uses, industrial processes, and feedstock users, and the establishment and implementation of priorities.

Natural Gas Wellhead Decontrol Act of 1989
(Public Law 101-60, 103 Statute 157)

This act amends the Natural Gas Policy Act of 1978 to eliminate wellhead price controls on the first sale of natural gas.

Natural Gas Pipeline Safety Acts

August 12, 1968, Public Law 90-481, 82 Statute 720

August 22, 1972, Public Law 92-401, 86 Statute 616

August 30, 1974, Public Law 93-403, 88 Statute 802

October 11, 1976, Public Law 94-447, 90 Statute 2073–2076

November 30, 1979, Public Law 96-129, 93 Statute 990–1003

January 14, 1983, Public Law 97-468, 96 Statute 2543

October 11, 1984, Public Law 98-464, 98 Statute 1821, 1823

April 7, 1986, Public Law 99-272, 100 Statute 139

October 22, 1986, Public Law 99-516, 100 Statute 2965

October 31, 1988, Public Law 100-561, 102 Statute 2806–2809, 2813, 2814

November 16, 1990, Public Law 101-599, 104 Statute 3038

Natural Gas Pipeline Safety Act of 1968
(Public Law 90-481, 82 Statute 720)

This act authorizes the secretary of transportation to prescribe safety standards for the transportation of natural and other gas by pipeline. Such standards must consider design, installation, construction, initial inspection and testing, operation, replacement, and maintenance of pipeline facilities.

Renewable Energy Legislation

Energy Security Acts

June 30, 1980, Public Law 96-294, 94 Statute 611

December 22, 1981, Public Law 97-98, 95 Statute 1299

Geothermal Energy Act of 1980
(Public Law 96-294, 94 Statute 763)

Solar Energy and Energy Conservation Act of 1980
(Public Law 96-294, 94 Statute 719)

Solar Energy and Energy Conservation Bank Act of 1980
(Public Law 96-294, 94 Statute 719)

November 30, 1983, Public Law 98-191, Title IV

October 17, 1984, Public Law 98-479, 98 Statute 2226

September 30, 1987, Public Law 100-122, 101 Statute 793

November 5, 1987, Public Law 100-154, 101 Statute 890

November 17, 1987, Public Law 100-170, 101 Statute 914

December 3, 1987, Public Law 100-179, 101 Statute 1018

December 21, 1987, Public Law 100-200, 101 Statute 1327

February 5, 1988, Public Law 100-242, 101 Statute 1950

Biomass Energy and Alcohol Fuels Act of 1980
(Public Law 96-294, 94 Statute 683)

August 13, 1981, Public Law 92-33, 95 Statute 622

April 15, 1985, Public Law 99-24, 99 Statute 50

April 7, 1986 , Public Law 99-272, 100 Statute 143

August 22, 1986, Public Law 99-386, 100 Statute 821

October 18, 1986, Public Law 99-500, 100 Statute 1783

October 30, 1986, Public Law 99-591, 100 Statute 3341

December 22, 1987, Public Law 100-202 , 101
Statute 1329

Renewable Energy Resources Act of 1980
(Public Law 96-294, 94 Statute 715)

Energy Security Act of 1980
(Public Law 96-294, 94 Statute 611)

This act was passed to guarantee the energy needs of the United States. It was based on the findings of Congress that:

1. The achievement of energy security for the United States is essential to the health of the national economy, the well-being of our citizens, and the maintenance of natural security
2. Dependence on foreign energy resources can be significantly reduced by the production from domestic resources of the equivalent of at least 500,000 barrels of crude oil per day of synthetic fuel by 1987 and of at least 2 million barrels of crude oil per day of synthetic fuel by 1992

3. Attainment of synthetic fuel production in the United States in a timely manner and in a manner consistent with the protection of the environment will require financial commitments beyond those expected to be forthcoming from nongovernmental capital sources and existing government incentives
4. Establishment of an independent federal entity of limited duration which will provide additional financial assistance in conjunction with private sources of capital to assist the development of domestic nonnuclear energy resources for the production of synthetic fuel which will facilitate the expeditious achievement of synthetic fuel production from domestic resources

The purpose of this act can be served by:

1. Demonstrating at the earliest feasible time the practicality of commercial production of synthetic fuel from domestic resources employing the widest diversity of feasible technologies
2. Fostering the creation of commercial synthetic fuel production facilities of diverse types with the aggregate capability to produce from domestic resources in an environmentally acceptable manner
3. Creating the United States Synthetic Fuels Corporation, a federal entity of limited duration formed to provide financial assistance to undertake synthetic fuel projects
4. Providing for financial assistance to encourage and ensure the flow of capital funds to those sectors of the national economy which are important to the domestic production of synthetic fuel
5. Encouraging private capital investment and activities in the development of domestic sources of synthetic fuel and to foster competition in the development of the nation's synthetic fuel resources
6. Encouraging and supplementing, but not competing with or supplementing private capital investments in the development of domestic sources of synthetic fuel
7. Fostering greater energy security and reducing the nation's economic vulnerability to disruptions in imported energy supplies

8. Giving special consideration to the production of synthetic fuel which has national defense applications and expediting its initial development through the Defense Production Act of 1950

Title I was devoted to the establishment of the United States Synthetic Fuels Corporation under the United States Fuel Corporation Act of 1980. Its primary purpose was to create the financial arrangements to develop the synthetic fuel industry.

Title II is cited as the Biomass Energy and Alcohol Fuels Act of 1980.

The purpose of this act was to develop:

A national program for increased production and use of biomass energy that does not impair the nation's ability to produce food and fiber on a sustainable basis for domestic and export use must be formulated and implemented within a multiple-use framework.

Title III considered Energy Targets.

Title IV is cited as the Renewable Energy Resources Act of 1980. The purpose of this act was:

To establish incentives for the use of renewable energy resources, to improve and coordinate the dissemination of information to the public with regard to renewable energy resources, to encourage the use of certain cost-effective solar energy systems and conservation measures by the federal government, to establish a program for the promotion of local energy self-sufficiency, to broaden the existing programs for accelerating the procurement and use of photovoltaic systems, and to provide further encouragement for the development of small hydroelectric power projects.

Title V is known as the Solar Energy and Energy Conservation Act of 1980 and the Solar Energy and Conservation Bank Act of 1980.

Title VI was intended as the Geothermal Energy Act of 1980. This act was based on Congress's finding that:

1. Domestic geothermal reserves can be developed into regionally significant energy sources providing for the economic health and national security of the nation
2. There are institutional and economic barriers to the commercialization of geothermal technology
3. Federal agencies should consider the use of geothermal energy in government buildings

Title VII considers Energy Conservation for Commercial Buildings and Multifamily Dwellings.

Title VIII amends the Strategic Petroleum Reserve portion of the Energy Policy and Conservation Act.

Regrettably, most of these acts were not funded, and implementation has been minimal.

Renewable Resources Extension Act

June 30, 1978, Public Law 95-306, 92 Statute 349

January 5, 1988, Public Law 100-231, 101 Statute 1565

November 28, 1990, Public Law 101-624, 104 Statute 3539

Renewable Energy and Efficient Technology Competitiveness Act of 1989 (Public Law 101-218, 103 Statute 1859)

This act provides federal assistance and leadership to a program of research, development, and demonstration of renewable energy and energy efficient technologies. The secretary of energy is empowered to develop this program to ensure a stable and secure future energy supply by:

1. Achieving, as soon as practicable, cost competitive use of those technologies without need of federal financial incentives
2. Establishing long-term federal research goals and multi-year funding levels
3. Directing the secretary of energy to undertake initiatives to improve the ability of the private sector to commercialize in the near term renewable energy and energy efficient technologies
4. Fostering collaborative research and development efforts involving the private sector through government support of a program of joint ventures

A number of specific programs include:

Wind The goals for the Wind Energy Research Program include improving design methodologies and developing more reliable and efficient wind turbines to increase the cost competitiveness of wind energy. Research efforts shall emphasize solving of near-term technical problems, developing technologies such as advanced airfoils and variable speed generators to increase wind turbine output, increasing the basic knowledge of aerodynamics, structural dynamics, and electrical systems, and improving the compatibility of electricity produced from wind fans with conventional utility processes.

Photovoltaics The goals of the Photovoltaic Energy Systems Program shall include improving the reliability and conversion efficiencies of, and lowering the costs of, photovoltaic conversion.

Solar Thermal Energy The goal of the Solar Thermal Energy Systems Program is to advance research and development to a point where solar thermal technology is cost competitive with conventional energy sources, and to provide the integration of this technology into the production of industrial process heat and the conventional utility network. Research and development shall emphasize development of a thermal storage technology to provide capacity for shifting power to periods of demand when full insolation is not available, improvement in receivers, energy conversion devices, and innovative concentrations using stretch membranes, levels, and other material; and exploration of advanced manufacturing techniques.

Other Renewable Technologies These include Biofuel Energy, Hydrogen Energy, Solar Building Energy, Ocean Energy, Geothermal Energy, Low-Head Hydro, and Energy Storage Systems.

The act encourages joint ventures to improve coordination in technology development among industries attempting to commercialize renewable energy. There is also the provision to establish a program to inform other countries of the benefits of these policies and to encourage the export of renewable industries technology.

Solar, Wind, Waste, and Geothermal Power Production Incentives Act of 1990 (November 15, 1990, Public Law 101-575, 104 Statute 2834)

This act encourages solar, wind, waste, and geothermal power production by removing the size limitations contained in the Public Utility Regulatory Policies Act of 1978.

Hydroelectrical Power Legislation

Federal Power Acts

June 10, 1920, Chapter 285, 41 Statute 1063

August 26, 1935, Chapter 687, 49 Statute 838

May 28, 1948, Chapter 351, 62 Statute 275

August 7, 1953, Chapter 343, 67 Statute 461

June 4, 1956, Chapter 351, 70 Statute 226

August 28, 1958, Public Law 85-791, 72 Statute 947

July 12, 1960, Public Law 86-619, 74 Statute 407

September 7, 1962, Public Law 87-647, 76 Statute 447

August 3, 1968, Public Law 90-451, 82 Statute 616

October 15, 1970, Public Law 91-452, 84 Statute 929

November 9, 1978, Public Law 95-617, 92 Statute 3134–3149

June 30, 1980, Public Law 96-294, 94 Statute 770

December 21, 1982, Public Law 97-375, 96 Statute 1826

October 16, 1986, Public Law 99-495, 100 Statute 1243–1249, 1252, 1255–1258

October 27, 1986, Public Law 99-546, 100 Statute 3056

October 6, 1988, Public Law 100-473, 102 Statute 2299

November 15, 1990, Public Law 101-575, 104 Statute 2834

Federal Water Power Acts

June 10, 1920, Chapter 285, 41 Statute 1063

March 3, 1921, Chapter 129, 41 Statute 1353

June 23, 1930, Chapter 572, 46 Statute 797

August 26, 1935, Chapter 687, 49 Statute 803

June 4, 1956, Chapter 351, 70 Statute 226

Federal Water Power Act
(Chapter 285, 41 Statute 1063)

This act established the Federal Power Commission. It was authorized to:

1. Make investigations and to collect and record data concerning the utilization of the water resources of the region to be developed, the waterpower industry and its relation to other industries and to interstate and foreign commerce, and covering the location, capacity, development costs, and relation to markets of power sites; and whether the power from government dams can be advantageously used by the United States for its public purposes, and what is the face value of such power
2. Make public, from time to time, the information secured
3. Issue licenses to citizens of the United States for the purpose of constructing, operating, and maintaining dams, water conduits, reservoirs, power houses, transmission lines, or other project works necessary or convenient for the development and improvement of navigation

Each applicant for a license had to conform to a number of requirements. These included:

1. The preparation of maps, plans, specifications, and estimates of cost as required for a full understanding of the proposed project
2. A comprehensive scheme of improvement and utilization for the purposes of navigation of waterpower development, and of other beneficial public uses
3. No substantial alteration or addition, not in conformity with the approved plans, shall be made to any dam or other project without approval of the commission
4. Annual payment had to be made by the licensee

This act initiated the federal government's involvement in the development and utilization of the water resources of the United States for production of electricity and navigability of rivers.

Energy and Water Development Appropriation Acts

September 25, 1970, Public Law 96-69, 93 Statute 437

October 1, 1980, Public Law 96-367, 94 Statute 1331

December 4, 1981, Public Law 97-88, 95 Statute 1135

December 22, 1981, Public Law 100-203, 101 Statute 1330-267

July 14, 1983, Public Law 98-50, 97 Statute 247

July 16, 1984, Public Law 98-360, 98 Statute 403

November 1, 1985, Public Law 99-141, 98 Statute 564

October 18, 1986, Public Law 99-500, 100 Statute 1783-213

October 30, 1986, Public Law 99-591, 100 Statute 3341-213

July 11, 1987, Public Law 100-71, 101 Statute 403

December 22, 1987 , Public Law 100-202, 101 Statute 1329-104

October 1, 1988, Public Law 100-463, 102 Statute 2270-45

November 17, 1988, Public Law 100-676, 102 Statute 4043

November 23, 1988, Public Law 100-707, 102 Statute 4715

September 29, 1989, Public Law 101-101, 103 Statute 657

November 5, 1990, Public Law 101-514, 104 Statute 2074

*Energy and Water Development Appropriation Act of 1970
(Public Law 96-69, 93 Statute 437)*

This act provides appropriation for energy and water development for the fiscal year ending September 30, 1980.

Under Title I the operating expenses for the Department of Energy are outlined:

Operating Expenses

Energy Supply, Research, and Development Activities	$2,048,523,000
Uranium Supply and Enrichment Activities	60,523,000
General Science and Research Activities	336,900,000

Atomic Energy Defense Activities	2,371,147,000
Departmental Administration	6,165,000
Plant and Capital Equipment	
Energy Supply, Research, and Development Activities	$448,478,000
Uranium Supply and Enrichment Activities	4,000,000
General Sciences and Research Activities	135,000,000
Atomic Energy Defense Activities	588,249,000
Department Administration	36,015,000
Power Marketing Administrations	
Operation and Maintenance, Alaska Power Administration	$2,660,000
Bonneville Power Administration Fund	1,000
Operation and Maintenance, Southeastern Power Administration	1,400,000
Operation and Maintenance, Southwestern Power Administration	32,180,000

Construction, Rehabilitation, Operation,
    and Maintenance, Western Area Power
    Administration
	122,800,000

Colorado River Basins Power
    Marketing Fund, Western Area Power
    Administration
	5,152,000

Emergency Fund, Western Area Power Administration	200,000
Federal Energy Regulatory Commission	67,187,000
Geothermal Resources Development Fuel	181,000

This act presents an example of the annul expenditures for development of the water resources of the nation.

Geothermal Energy Legislation

Geothermal Steam Acts

December 24, 1970, Public Law 91-581, 84 Statute 1566

September 22, 1988, Public Law 100-443, 102 Statute 1766

Geothermal Steam Act of 1970
(Public Law 91-581, 84 Statute 1566)

This act authorizes the secretary of the interior to develop geothermal steam and associated geothermal resources. A number of provisions are:

1. If lands are to be leased under this act and are within any known geothermal resource area, they shall be leased to the highest responsible qualified bidder by competitive bidding. Other areas may be leased without competitive bidding.
2. Mining claims are limited to 10,240 acres.
3. A royalty of not less than 10 percent or more than 15 percent of the amount or value of steam, or any other form of energy, will be paid.
4. A royalty of not more than 5 percent of the value of any by-product derived from steam production will be paid.
5. Geothermal leases shall have a primary term of 10 years, and no longer than 40 years, before renewal.

Geothermal Steam Act Amendment of 1988

September 22, 1988, Public Law 100-443, 102 Statute 1766

Geothermal Energy Research, Development, and Demonstration Acts

September 3, 1974, Public Law 93-410, 89 Statute 1079

February 25, 1978, Public Law 95-238, 92 Statute 86-89

Geothermal Energy Acts

June 30, 1980, Public Law 96-294, 94 Statute 768

Solar Energy Legislation

Solar Energy Research, Development, and Demonstration Acts

October 28, 1974, Public Law 93-473, 88 Statute 1431

October 19, 1980, Public Law 96-470, 94 Statute 2243

Solar Energy Research, Development, and Demonstration Act of 1974 (Public Law 93-473, 88 Statute 1431)

This act authorizes a vigorous federal program of research, development, and demonstration to ensure the utilization of solar energy as a viable source for our material energy needs.

The act was based on congressional findings that:

1. The needs of a viable society depend on an ample supply of energy
2. The current imbalance between domestic supply and demand for energy is likely to be present for some time
3. Dependence on nonrenewable energy resources cannot be continued indefinitely, particularly at current rates of consumption
4. It is in the nation's interest to expedite the long-term development of renewable and nonpolluting energy resources, such as solar energy
5. The various solar energy technologies are today at widely differing stages of development, with some already near the stage of commercial application and others still requiring basic research
6. The early development and export of viable equipment utilizing solar energy, consistent with the established preeminence of the United States in the field of high technology products, can make a valuable contribution to our balance of trade
7. The mass production and use of equipment utilizing solar energy will help to eliminate the dependence of the United States upon foreign energy sources and promote national defense

This act establishes the Solar Energy Coordination and Management Project. This project has the overall responsibility for the provision of effective management and coordination with respect to a national solar energy, research, development, and demonstration project, including:

1. The demonstration and evaluation of this resource base, including temporal and geographic characteristics
2. Research and development of solar technologies
3. The demonstration of appropriate solar energy technologies

Solar Heating and Cooling Demonstration Act

1974, Public Law 93-409, 88 Statute 1069

*Solar Heating and Cooling Demonstration Act of 1974
(Public Law 93-409, 88 Statute 1069)*

This act provided for the early development and commercial demonstration of the technology of solar heating and/or combined solar heating and cooling systems.

In passing this act Congress found that:

1. The current imbalance between supply and demand for fuels and energy is likely to be present for some time
2. The early demonstration of the feasibility of using solar energy for the heating and cooling of buildings could help to relieve the demand upon present fuels and energy supplies
3. The technologies for solar heating are close to the point of commercial application in the United States

The act stated that the policy of the United States was to demonstrate within a three-year period the practical use of solar heating technology.

Demonstrations were to be conducted on:

1. Heating systems to be used in residential dwellings
2. Heating and cooling systems for commercial and residential buildings

Solar Photovoltaic Energy Research, Development, and Demonstration Act

1978, Public Law 95-590, 92 Statute 2513

*Solar Photovoltaic Energy Research, Development, and
Demonstration Act of 1978 (Public Law 95-590, 92 Statute 2513)*

This act provides legislation for an accelerated program of research, development, and demonstration of solar photovoltaic energy technologies leading to early competitive commercial applicability of such technologies to be carried out by the Department of Energy, with the support of the National Aeronautics and Space Administration, the National Bureau of Standards, the General Services Administration, and other federal agencies.

The act was prompted by the growing energy problem. Congress thereby found that:

1. The early development and widespread utilization of photovoltaic energy systems could significantly expand the domestic energy resource base of the United States, thus lessening its dependence on foreign supplies
2. The establishment of sizable markets for photovoltaic energy systems will justify private investment in plant and equipment necessary to realize the economies of scale, and will result in significant reductions in the unit costs of these systems
3. The use of solar photovoltaic energy systems for certain limited applications has already proved feasible
4. There appear to be no technical obstacles to the widespread commercial use of solar photovoltaic energy technologies
5. An aggressive research and development program should solve existing technical problems of solar photovoltaic systems and, supported by an assured and growing market for photovoltaic systems during the next decade, should maximize the future contribution of solar photovoltaic energy to the nation's future energy production
6. It is the proper and appropriate role of the federal government to undertake research and development and demonstration programs in solar photovoltaic energy technologies, and to supplement and assist private industry and other entities and thereby the general public, so as to hasten the general commercial use of such technologies
7. The high cost of imported energy sources impairs the economic growth of many nations which lack sizable domestic energy supplies and are unable to develop their resources
8. Photovoltaic energy systems are economically competitive with conventional energy resources for a wide variety of applications in many foreign nations at

the present time, and will find additional applications
with continued cost reduction

9. The early development and export of solar photovoltaic energy systems, consistent with the established
preeminence of the United States in the field of high
technology products, can make a valuable contribution
to the well-being of the people of other nations and the
nation's balance of trade

10. The widespread use of solar photovoltaic energy
systems to supplement and replace conventional
methods for the generation of electricity would have
a beneficial effect upon the environment

11. To increase the potential use of solar photovoltaic
energy systems in remote locations, programs leading to the development of inexpensive and reliable
systems for the storage of electricity should be
pursued

The objectives of this act were to:

1. Double the production of solar photovoltaic energy
systems each year during the decade starting with the
fiscal year 1979, so as to reach a photovoltaic energy
production of approximately two million peak kilowatts
and a total cumulative production of several systems of
approximately five million peak kilowatts by fiscal year
1988

2. Reduce the average cost of installed solar photovoltaic energy systems to \$1 per peak watt by fiscal
year 1988

3. Stimulate the purchase by private buyers of at least 90
percent of all photovoltaic systems produced in the
United States during fiscal year 1988

The secretary of energy was to implement the program by ac-
celeration of research, development, and demonstration. A Solar
Photovoltaic Advisory Committee was nominated to advise the
secretary of energy on all aspects of the project. The secretary was
authorized to encourage to the maximum extent practicable, international participation and cooperation in the development and
utilization of photovoltaic systems.

Wind Energy Legislation

Wind Energy Systems Acts

September 8, 1980, Public Law 96-345, 94 Statute 1139

August 26, 1986, Public Law 99-386, 100 Statute 821

Wind Energy Systems Act of 1980
(Public Law 96-345, 94 Statute 1139)

This act provided for an accelerated program of wide energy research development and demonstration, to be carried out by the Department of Energy and with the support of the National Aeronautics and Space Administration and other federal agencies.

Congress based passage of this act on:

1. The nation's best interests being served by providing opportunities for the increased production of electricity from renewable energy sources
2. The early widespread utilization of wind energy for the generation of electricity and for mechanical power could lead to relief on the demand for existing nonrenewable fuel and energy supplies
3. The use of large wind energy systems for certain limited applications is already economically feasible
4. An aggressive research, development, and demonstration program to accelerate widespread utilization of wind energy should solve existing technical problems of converting wind energy into electricity and mechanical power
5. It is the role of the federal government to undertake research and development, to participate in demonstration programs for wind energy systems, and to assist private industry and the general public in fostering the widespread utilization of such systems
6. The widespread use of wind energy systems will have a beneficial effect on the environment
7. The evaluation of the performance and reliability of wind energy technologies can be expedited by the testing of prototypes under carefully controlled conditions

This act declared the policy of the United States to:

1. Reduce the average cost of electricity produced by installed wind energy systems by the end of fiscal year 1988 to a level competitive with conventional energy sources
2. Reach a total megawatt capacity in the United States from wind energy systems, by the end of fiscal year 1988, of at least 800 megawatts, of which at least 100 megawatts are provided by small wind energy systems
3. Accelerate the growth of a commercially viable and competitive industry to make wind energy systems available to the general public as an option in order to reduce national consumption of fossil fuels

Under the act the secretary of energy was to implement the program. As part of the project the secretary was to initiate a three-year national wind resources assessment program. The program was to include:

1. Development of activities to validate existing assessment of known wind resources
2. Performance of wind resource assessments in regions of the United States where the use of wind energy may prove feasible
3. Initiation of a general site prospecting program
4. Establishment of standard wind data collection and siting techniques
5. Establishment, with other agencies, of a national wind data center which shall make public information available on the known wind energy resources of various segments throughout the United States

As of 1992 little had been accomplished to implement this legislation.

Energy and the Environment Legislation

Clean Air Acts

July 14, 1955, Chapter 360, 69 Statute 322

December 17, 1963, Public Law 88-206, 77 Statute 392

October 20, 1965, Public Law 89-272, 79 Statute 992

October 15, 1966, Public Law 89-675, 80 Statute 954

November 21, 1967, Public Law 90-148, 81 Statute 485

December 5, 1969, Public Law 91-137, 83 Statute 283

December 31, 1970, Public Law 91-604, 84 Statute
1676–1713

November 18, 1971, Public Law 92-157, 85 Statute 464

April 9, 1973, Public Law 93-15, 87 Statute 11

June 22, 1974, Public Law 93-319, 88 Statute 248–259,
261, 265

August 7, 1977, Public Law 95-95, 91 Statute 685–796

November 16, 1977, Public Law 95-190, 91 Statute
1399–1404

July 2, 1980, Public Law 96-300, 94 Statute 831

July 17, 1981, Public Law 97-23, 95 Statute 139

December 8, 1983, Public Law 98-213, 97 Statute 1461

November 15, 1990, Public Law 101-549, 104 Statute 2399

By the 1950s it had become evident that air pollution was not a local or even a regional phenomenon but a national problem, so that local and state regulations could not be completely effective in dealing with it. The first federal attempts to provide a framework for the control of air pollution were tentative, but the legislation and regulatory acts between 1955 and 1970 gradually began to address the problem more directly.

In 1965 a Special Subcommittee on Air and Water Pollution of the Department of Health, Education, and Welfare testified:

> Serious air pollution problems arise from the ever-increasing use of motor vehicles, [and] our rising demand for energy derived from burning of sulfur-bearing fuels. The national importance of resolving these problems are beyond dispute. They are among the most significant factors in the growing and worsening air pollution problem currently faced by thousands of American communities.

None of these problems is technically insurmountable. Our current technical capacity varies and even where our knowledge and skills are less refined, there exists partial relations. Although Clean Air Acts are more than 25 years old, a majority of Americans are still breathing air that does not meet the federal health-based air quality standards.

Researchers have demonstrated adverse effects at levels of exposure previously presumed to be safe. In some areas, for example, residents who do not smoke may have damaged lungs as unhealthy as the lungs of heavy smokers, due to exposure to air pollution. We are continuously exposed to thousands of different air pollutants emitted every day. Exposure to these mixes of pollutants can produce more adverse health effects than exposure to many industrial pollutants. The synergistic effects must be considered if public health is to be protected.

Clean Air Act Amendment of 1970
(Public Law 91-604, 84 Statute 1676-1713)

The Clean Air Act Amendment of 1970 inaugurated the modern era of air pollution control. This act recognized that air pollution was a national problem and could only be addressed through federal programs. Thus, this act plus those that followed in 1977, 1980, 1981, 1983, and 1990 established a partnership between the states and the federal government. The Environmental Protection Agency set national quality standards and the states were responsible for meeting them.

In turn, the states had to develop State Implementation Plans (SIP) and submit them to the EPA for review. By this procedure the federal government controlled the states' efforts:

The major provisions of the act included:

1. An expanded research program with special emphasis to be placed on:
   a. The means to clean fuels before combustion
   b. The development of new and synthetic fuels
   c. Epidemiological studies of the effects of air pollutants on mortality and morbidity
   d. Studying the immunological, biochemical, physiological, and toxicological effects of air pollutants
2. An assessment of the causes and effects of noise pollution
3. Establishment of primary (health) and secondary (welfare) ambient air quality standards

4. Emission limitations enforceable by both state and federal governments
5. Transportation plans that must include the "land use and transportation controls" necessary to attain air quality standards

When the Clean Air Amendments were passed in 1970, Congress set 1975 as the deadline for meeting the primary air quality standards. By 1977, two years beyond the original deadline, 78 urban areas had not attained the ozone standards established by the EPA. In 1977 the deadline for meeting the ozone and carbon monoxide standards was moved to 1982, and in 1982 the deadline was extended to 1987 for those areas still in violation. Yet in 1989, over 150 million people lived in an area which exceeded one or both of the standards.

The reasons for the slow progress to achieve healthy air in most urban areas are many. Among these are:

1. The understatement of emissions in initial plans (SIPs) submitted by the states and approved by the EPA
2. Inadequacies in models used to predict ambient air quality
3. Failure of the state to implement some of the controls committed to in their SIPs
4. Failure of some of the controls to achieve the projected emission reductions
5. Failure of the EPA and the states to require additional controls

In 1989 it was evident that the Clean Air Act required a major revision. This resulted in the 1990 Clean Air Act Amendments.

Clean Air Act, Amendments of 1990 (Public Law 101-549, Statute 1630)

The 1990 amendments to the Clean Air Act established more stringent standards for the attainment and maintenance of ambient air quality.

Title I of the act revises standards for the attainment and maintenance of national ambient air quality. Emphasis is placed on ozone, carbon monoxide, particle matter, sulfur oxides, nitrogen dioxide, and lead nonattainment areas. To implement these controls the federal and state governments are to cooperate in developing attainment plans.

Title II of the act considers provisions relating to mobile sources. A wide range of topics include control of vehicle refueling emissions, emission standards, mobile source-related air toxins, state fuel regulations, reformulated gasoline and oxygenated gasoline, lead phasedown, heavy-duty trucks, carbon buses, clean-fuel vehicles, motor vehicle testing and certification, carbon monoxide emission at cold temperatures, and others.

Title III is devoted to hazardous air pollutants. These topics include rich assessment and chemical process safety management, solid waste combustion, and ash management and disposal.

Title IV considers acid deposition control. Such factors are treated as fossil fuel use, acid deposition standards, national acid lakes registry, industrial SO_2 emissions, monitoring the acid rain program in Canada, clean coal technologies export programs, acid deposition research by the United States Fish and Wildlife Service, study of buffering and neutralizing agents, and special clean coal technology projects. The goal of the program is to reduce the adverse effects of ten million tons of sulfur dioxide and two million tons of nitrogen dioxide emission into the atmosphere. These reductions are to be achieved by requiring compliance with prescribed emission limitations by specified deadlines. It is also the purpose of this title to encourage energy conservation, use of renewable and clean alternative technologies, and pollution prevention as a long-range strategy for reducing air pollution and other adverse impacts of energy production and use.

Title V considers permit programs, application requirements, and conditions.

Title VI treats stratospheric ozone protection. Sections are devoted to monitoring and reporting requirements, phaseout of production and consumption of chlorofluorocarbons, accelerated schedule, exchange, national recycling and emission reduction programs, servicing of motor vehicle air conditioners, stopping production of nonessential products containing chlorofluorocarbons, developing safe alternative policies, encouraging passage of the Montreal Protocol, and increasing international cooperation.

Title VII describes provisions relating to enforcement and Title VIII is related to miscellaneous provisions. Title IX strengthens clean air research. Major problems to be considered are individual and complex mixtures of air pollutants, the establishment of a national network to monitor, collect, and compile data on the status and trends of air emissions, deposition, air quality, surface water quality, forest conditions, visibility impairment, and the development of improved methods and technologies for sampling measurement, monitoring, analysis, and modeling to increase understanding of the sources of ozone, ozone formation, ozone transport, regional influences on urban ozone, regional ozone trends, and the interaction of ozone with other pollutants.

The last two titles consider disadvantaged business concerns and clean air employment transmission assistance.

Although there is recognition that clean air is a necessary prerequisite for the health and welfare of the nation, 40 years of local, state, and federal legislation have not solved the problem, and many complex issues remain unresolved. These include such questions as: What is the economic cost of clean air? Can the present economy be maintained and also achieve clean air as well as solving the other environmental problems of the earth? What are the social costs to secure clean air? If clean air is not achieved, what will be the health costs to the individual and to the nation? How will traditional patterns of life be changed? Can the 1990 act be any more effective than previous legislation? These are only a few of the questions that must be answered in the immediate future if a clean environment is to be secured.

Coal Mining Legislation

Surface Mining Control and Reclamation Acts

August 3, 1977, Public Law 95-87, 91 Statute 445

March 7, 1978, Public Law 95-240, 92 Statute 109

August 11, 1978, Public Law 95-343, 92 Statute 473

November 9, 1978, Public Law 95-617, 92 Statute 3166, 3167

December 11, 1980, Public Law 96-511, 94 Statute 2826

December 22, 1980, Public Law 97-98, 95 Statute 1344

October 12, 1984, Public Law 98-473, 98 Statute 1875

October 18, 1986, Public Law 99-500, 100 Statute 1783–267

October 30, 1986, Public Law 99-591, 100 Statute 3341–267

May 7, 1987, Public Law 100-34, 101 Statute 300

July 11, 1987, Public Law 100-71, 101 Statute 416

November 5, 1990, Public Law 101-508, 104 Statute 1388–289

In 1977 Congress enacted the Surface Mining Control and Reclamation Act (SMCRA) to protect the environment from the adverse effects of surface mining. In passing this law Congress found that:

1. Extraction of coal and other minerals from the earth can be accomplished by various mining methods including surface mining
2. Many surface mining operations result in disturbances of surface areas that burden and adversely affect commerce and the public welfare by destroying or diminishing the utility of land for commercial, industrial, residential, recreational, agricultural, and forestry purposes, by causing erosion and landslides, by contributing to floods, by polluting the water, by destroying fish and wildlife habitats, by impairing natural beauty, by damaging the property of citizens, by creating hazards dangerous to life and property, by degrading the quality of life in local communities, and by counteracting governmental programs and efforts to conserve soil, water, and other natural resources
3. Surface mining and reclamation technology is now developed so that effective and desirable regulation of surface coal-mining operations by the state and the federal governments is an appropriate and necessary means to minimize the adverse social, economic, and environmental effects of such mining operations
4. Because of the diversity in terrain, climate, biological, chemical, and other physical conditions in areas subject to mining operations, the primary governmental responsibility for developing, authorizing, issuing, and enforcing regulations for surface mining and reclamation operations should rest with the states

5. There are a substantial number of areas of land throughout major regions of the United States disturbed by surface and underground mining on which little or no reclamation was conducted, and the impact from these unreclaimed lands imposes social and economic costs on residents in nearby and adjoining areas as well as continuing to impair environmental quality

To administer the act, an Office of Surface Mining Reclamation and Enforcement was established in the Department of the Interior. Within the states, Mining and Mineral Resources and Research Institutes were established to provide reclamation information. A trust fund was created to pay for reclaiming abandoned strip-mined land.

4

Directory of Organizations

ORGANIZATIONS THAT CONSIDER the development and utilization of energy are grouped into five major categories. The first considers the principal U.S. government agencies. These agencies play a prominent role in establishing energy policies in the nation. The second group lists many intergovernmental advisory committees that have specific and limited objectives. Many provide information to governmental agencies and private organizations. The third group, private organizations, have usually been developed to provide advice on specific problems such as energy conservation and its effective utilization. Most of these groups were founded after the energy crisis of the 1970s. The fourth group of organizations include the individual energy sources. Finally, the fifth group lists international organizations that reflect the worldwide scope of the energy problem.

U.S. Government Organizations

Department of Energy
1000 Independence Avenue, SW
Washington, DC 20585

Description: Established October 1, 1977, to coordinate major federal energy functions.

Purpose: The basic functions of the Department of Energy (DOE) are the overall planning, direction, and control of the energy policies of the nation.

Activities: The Department of Energy has a number of offices and programs that carry out specific functions, such as:

ENERGY RESEARCH   The Office of Energy Research advises the Department of Energy secretary on the DOE's physical and energy research and development programs, the use of multipurpose laboratories, education, training for basic and applied research, and financial assistance and budgetary priorities for these activities.

This office manages the basic energy sciences, high-energy physics, and fission energy research programs; administers DOE programs supporting university researchers; funds research in mathematical and computational sciences critical to the use and development of supercomputers; and administers a financial support program for research and development projects not funded elsewhere in the department. The office also manages a research program aimed at determining the generic environmental, health, and safety aspects of energy technologies and programs.

The office monitors DOE research and development programs for deficiencies or duplication of effort and, in conjunction with the assistant secretary for International Affairs and Energy Emergencies, monitors the international exchange of scientific and technical personnel.

FOSSIL ENERGY   The assistant secretary for fossil energy is responsible for research and development programs involving fossil fuels—coal, petroleum, and gas. The fossil energy program involves applied research, exploratory development, and limited proof-of-concept testing targeted to high-risk and high-payoff endeavors. The objective of the program is to provide the general technology and knowledge base that the private sector can use to complete development and initiate commercialization of advanced processes and energy systems. The program is principally executed through two energy technology centers located in the field.

The assistant secretary also manages the Clean Coal Technology Program, the Strategic Petroleum Reserve, the Naval Petroleum and Oil Shale Reserves, and the Liquified Gaseous Fuels Spill Test Facility.

CONSERVATION AND RENEWABLE ENERGY   The assistant secretary for conservation and renewable energy is responsible for formulating and directing programs designed to increase the production and utilization of renewable energy (solar, biomass, wind, geothermal, alcohol fuels, etc.) and improving the energy efficiency of transportation, buildings, industrial systems and related processes through support of long-term, high-risk research and development activities.

The assistant secretary also has responsibility for administering statutorily mandated assistance programs that provide financial assistance for state energy planning, weatherization of housing owned by the poor and disadvantaged, and the implementation of energy conservation measures by schools and hospitals, local units of government, and public care institutions. In addition, the office coordinates and oversees the operations of three power-marketing administrations.

INTERNATIONAL AFFAIRS AND ENERGY EMERGENCIES   The assistant secretary of international affairs and energy emergencies develops, manages, and directs programs and activities relating to the international aspects of overall energy policy; ensures that U.S. international energy policies and programs conform to national goals, legislation, and treaty obligations; advises the secretary on international energy negotiations; assesses world price and supply trends; and coordinates cooperative international energy programs with foreign governments and international organizations such as the International Energy Agency and the International Atomic Energy Agency.

The assistant secretary also directs and coordinates the department's energy emergency preparedness and energy emergency operations efforts, develops contingency plans and tests and evaluates response plans, and develops and maintains continuity of government and national emergency plans.

NUCLEAR ENERGY   The assistant secretary of nuclear energy administers the department's research and development programs associated with fission energy. This includes programs relating to nuclear reactor development, both civilian and naval, nuclear fuel cycle, space nuclear applications, and uranium enrichment.

The assistant secretary also manages the department's Remedial Action Program to treat and stabilize radioactive wastes and perform decontamination and decommissioning at DOE surplus sites. In addition, the assistant secretary conducts technical analysis and provides advice concerning nonproliferation; assesses alternative nuclear systems and new reactor and fuel cycle concepts; and evaluates proposed advanced nuclear fission energy concepts and technical improvements for possible application to nuclear power plant systems.

ENERGY INFORMATION ADMINISTRATION   The Energy Information Administration is responsible for the timely and accurate collection, processing, and publication of data in the areas of energy resource reserves, energy production, demand, consumption, distribution, and technology.

FEDERAL ENERGY REGULATORY COMMISSION An independent, five-member commission within the Department of Energy, the Federal Energy Regulatory Commission has retained many of the functions of the Federal Power Commission, such as setting rates and charges for the transmission and sale of electricity and the licensing of hydroelectric power projects. In addition, the commission establishes rates or charges for the transportation of oil by pipeline, as well as the valuation of such pipelines.

DOE FIELD STRUCTURE The vast majority of the department's energy research and development, uranium enrichment, nuclear weapons research development, and testing and production activities are carried out by contractors who operate government owned facilities. To develop this program the Department of Energy has eight operations offices.

REGULATORY PROGRAMS The Department of Energy's enforcement program is conducted by the field staff under the supervision and direction of the Economic Regulatory Administration.

POWER ADMINISTRATION The power administrations include:

Bonneville Power Administration

Southeastern Power Administration

Alaska Power Administration

Southwestern Power Administration

Western Area Power Administration

These administrations market power to the regions at the lowest rates. Power is sold at wholesale to utilities and directly to industries. The administration also disposes of surplus electric power and energy generated at federal reservoir projects. The programs of the power administrations include the negotiation, preparation, execution, and administration of contracts for the distribution of electric power; the preparation of wholesale rates and repayment schedules; the provision by construction, contract, or otherwise, of transmission and related facilities to interconnect reservoir projects and to service contractual loads; and activities pertaining to the planning and operation of power facilities.

Nuclear Regulatory Commission
Washington, DC 20555

Description: Established as an independent regulatory agency under the provisions of the Energy Reorganization Act of 1974 and Executive Order 11834 of January 15, 1975, the commission is delegated all licensing and related regulatory functions formerly assigned to the Atomic Energy Commission, which was established by the Atomic Energy Act of 1946, and as amended in 1954.

Purpose: Ensures that civilian users of nuclear materials and facilities operate in a manner consistent with public health and safety, environmental quality, national security, and antitrust laws. The major share of the commission's effort is focused on regulating the use of nuclear energy to generate electric power.

Activities: Fulfills its responsibilities through a system of licensing and regulation that includes:

1. Licensing the construction and generation of nuclear reactors and other nuclear facilities and the possession, use, processing, handling, and disposal of nuclear materials.
2. Regulation of licensed activities including assurance that measures are taken for the physical protection of facilities and materials.
3. Development and implementation of rules and regulations governing licensed nuclear activities.
4. Inspection of licensed facilities and activities.
5. Investigation of nuclear incidents and allegations concerning any matter regulated by NRC.
6. Enforcement of NRC licenses and regulations by the issuance of orders, civil penalties, and other types of actions.
7. Conduct of public hearings on nuclear and radiological safety, environmental concerns, common defense and security, and antitrust matters.
8. Development of effective working relationships with states regarding the regulation of nuclear materials.

The commission guarantees that adequate regulatory programs are maintained by those states that exercise, by agreement with the commission, regulatory control over certain nuclear materials within their respective borders. In addition, a systematic revision of operational data, including reports of accidents and other events from nuclear power plants, is performed in order to detect trends that will better enable the agency to forecast and solve safety problems. Inspection of commission licensed activities is carried out from five regional offices.

Publications: Annual Report, provides a survey of major agency activities for the year; *Nuclear Regulatory Commission Issuances* (monthly, quarterly and semiannual indexes), a compilation of adjudications and other issuances of the commission. Other publications include: *Licensed Operating Reactors—Status Summary Report* (monthly), *Licensed Contractor and Vendor Inspection Status Report* (quarterly), *Report to Congress on Abnormal Occurrences* (quarterly), *Regulatory and Technical Reports* (quarterly), *Nuclear Commission Rules and Regulations* (monthly), *Information Report on State Legislation* (monthly), and Weekly Information Report.

The commission produces a variety of scientific, technical, and administrative information publications dealing with licensing and regulating civilian nuclear power. Information Agency publications can be obtained from, *Title List of Documents made Publicly Available,* NUREG-0540. The commission maintains its principal Public Document Room at 2120 L Street, NW, Washington, D.C., 20555. Members of the public may visit the Public Document Room and examine any document available there, or write or phone in requests (202/634-3273).

U.S. Council for Energy Awareness (USCEA)
1776 I Street, NW, Suite 400
Washington, DC 20006

Description: Founded in 1987, the council has 405 members and a staff of 70, with a $18,000,000 budget. Members are interested in energy matters.

Purpose: Promote the development and use of nuclear energy.

Activities: Engaged in development and use of nuclear energy, especially nuclear-produced electricity and other energy matters. Bestows awards, including the annual Forum Award, for contributions to the field of nuclear energy; maintains a speakers' bureau; compiles statistics and data; and has a library. Three to four conferences are held a year.

Publications: Annual Report, INFO (monthly), *Nuclear Industry* (quarterly), and reports related to nuclear energy.

U.S. Intergovernmental Advisory Committees

Advisory Committee on Renewable Energy and
Energy Efficiency Joint Ventures
Department of Energy
Forrestal Building, 6C-016
1000 Independence Avenue, SW
Washington, DC 20585

Description: Established by the secretary of energy under authority of Public Law 101-218, the Renewable Energy and Energy Efficiency Technology Competitiveness Act of 1989, dated December 11, 1989. Members are appointed by the secretary of energy. Membership includes at least one representative from each of the following areas: Secretary of commerce, National Laboratories of the Department of Energy, Solar Energy Research Institute, Electric Power Research Institute, Gas Research Institute, National Institute of Building Sciences, National Institute of Standards and Technology, firms in renewable energy manufacturing industries, and firms in major energy efficiency manufacturing industries.

Purpose: Advisory committee to the Department of Energy.

Activities: Advises the secretary of energy on developments for joint ventures in research and development of renewable energy and energy efficient technologies such as photovoltaics, wind energy, and solar thermal energy. It also advises the secretary on the implementation of the joint venture program.

Publications: None.

Basic Energy Sciences Advisory Committee (BESAC)
Office of Energy Research
Department of Energy
Washington, DC 20585

Description: Established September 30, 1986, as a public advisory committee. There are 16 to 19 members, appointed by the secretary of energy, who serve one-year terms. Support services are provided by the Office of Energy Research.

Purpose: Acts in a public advisory capacity.

Activities: Advises the secretary of energy on the Basic Energy Sciences Program that was established as a means to pursue long-term research to build a base of scientific and engineering knowledge to be used by the federal government in developing new concepts and processes for energy production, conversion, and use. The committee reviews the program's work and then makes recommendations. The committee meets four times a year, at least once each quarter.

Publications: Review of the Basic Energy Sciences Program of the Department of Energy (annual), *High-Tc Superconducting Magnet Applications in Particle Physics* (1987), and *Joint BESAC/HERAC Report on Global Change* (1989).

Committee on Renewable Energy, Commerce and Trade (CORECT)
Office of the Secretary
Department of Energy
1000 Independence Avenue, SW
Washington, DC 20595

Description: An interagency working committee of the Department of Energy established in 1984. There are 14 members appointed by the secretary of energy.

Purpose: Advises the secretary of energy.

Activities: Makes recommendations and provides advice to the secretary on the use of renewable energy products and services. It works to establish a government-industry plan to increase the U.S. market share in the trade of renewable energy technologies and products.

Publications: None.

Interagency Energy Management Task Force
Office of the Secretary
Department of Energy
1000 Independence Avenue, SW
Washington, DC 20585

Description: Established by the secretary of energy under authority of Public Law 100-615, the Federal Management Improvement Act of 1988. Members are chief energy managers of federal departments and agencies.

Purpose: Disseminate information on energy developments.

Activities: Acts as an interagency task force for the Department of Energy. It collects and disseminates energy information to federal agencies, state, and local governments. It studies the use of cogeneration facilities, renewable resources, methods to conserve energy, and promotes the efficient use of energy. The task force meets at the request of the secretary of energy, usually twice a year.

Publications: None.

National Energy Extension Service Advisory Board (NEESAB)
Office of State and Local Assistance Programs
Department of Energy
1000 Independence Avenue, SW
Washington, DC 20585

Description: Established June 14, 1978, by Title V of Public Law 95-39, the National Energy Extension Service Act. It has no more than 20 members, appointed by the secretary of energy, and selected from persons representative of state, county, and local governments, universities, small business, and agriculture.

Purpose: Advisory.

Activities: Founded to advise the director of energy extension, secretary of the Department of Energy, and Congress on the effectiveness of the

Extension Service Program, and the plans of governors of each state as they are implemented to achieve the goals of the National Energy Extension Service Act. Meets three times a year.

Publications: Annual Report.

National Petroleum Council (NPC)
1625 K Street, NW
Washington, DC 20006

Description: Established June 18, 1946, by the secretary of the interior at the request of President Truman. NPC has about 150 members who serve one-year terms. Support services are provided by the Office of Fossil Energy.

Purpose: Transferred to the Department of Energy in 1977 to serve as an industry advisory council.

Activities: Advise, inform, and make recommendations to the secretary of energy on matters relating to oil and gas or oil and gas industries. Matters considered include a wide range of technical and policy questions. It meets at least twice a year.

Publications: A number of reports each year.

Pacific Northwest Electric Power and Conservation Planning Council
851 SW Sixth Avenue, Suite 1100
Portland, OR 97204

Description: Established in 1981 pursuant to Public Law 96-501, the Pacific Northwest Electric Power Planning and Conservation Act. It is composed of two people from each state in the Pacific Northwest Region. It is an interstate compact agency and is also known as the Northwest Power Planning Council.

Purpose: Prepare and adopt a regional conservation and electric power plan.

Activities: The council protects and enhances fish and wildlife in the Columbia River Basin; assists electrical consumers through the use of the Federal Columbia River Power System to achieve cost-effective energy conservation, encourages development of renewable energy resources, assures the region of efficient and adequate power supply, and restores fish and wildlife harmed by hydropower development. The council has several subsidiary units. Meets once a month.

Publications: None.

Research Development and Demonstration Task Force
851 SW Sixth Avenue, Suite 1100
Portland, OR 97204

Description: Established September 17, 1986, by the Pacific Northwest Electric Power and Conservation Planning Council as authorized by Public Law 96-501.

Purpose: Acts as an advisory force for the council.

Activities: Make recommendations to the council with respect to the best allocations of regional funding for research and development, coordinate the research and development to confirm the cost effectiveness of conservation and renewable, high-efficiency resources.

Publications: None.

Secretary of Energy Advisory Board (SEAB)
Forrestal Building, Room 7B-198/FORS
1000 Independence Avenue, SW
Washington, DC 10585

Description: Established January 2, 1990, it serves as a public advisory board to the Department of Energy. Thirty to 35 members, appointed by the secretary of energy, are professionals in the sciences and various aspects of energy from various geographic regions.

Purpose: Advise the secretary of energy.

Activities: Advises and gives guidance to the secretary of energy on research, development, energy, national defense responsibilities, and the activities of the Department of Energy. Meets twice a year.

Publications: None.

State Agency Advisory Committee
851 SW Sixth Avenue, Suite 1100
Portland, OR 97204

Description: Established June 7, 1984, by the Pacific Northwest Electric Power and Conservation Planning Council, as authorized by Public Law 501, the Pacific Northwest Electric Power Planning and Conservation Act. It is also known as the State Agency Advisory Task Force.

Purpose: Advises the public about the work of the council.

Activities: Provides a means for the exchange of ideas and information between the council and state agencies. The council's 1986 plan includes the evaluation of resource costs and benefits and developing a resource plan that minimizes costs to the region. Views of the task force and council may differ, so this gives an opportunity for both to express their views.

Publications: None.

State Energy Advisory Board

Office of the Secretary
Department of Energy
Forrestal Building, 7A-257
1000 Independence Avenue, SW
Washington, DC 20585

Description: Established by the secretary of energy under authority of Section 365 of the Energy Policy and Conservation Act (42USC6325) as amended by Section 5 of Public Law 101-440, the Energy Efficiency Programs Improvement Act of 1990, October 18, 1990. Members include 18 to 21 persons, mostly state employees responsible for developing state energy conservation plans, such as low-income weatherization assistance programs and renewable energy programs. Of the original board, one-third serve one-year terms, one-third serve two-year terms, and one-third serve three-year terms. Other members serve three-year terms.

Purpose: Advisory capacity.

Activities: Makes recommendations to the assistant secretary of Conservation and Renewable Energy on energy efficiency goals and the purposes of programs carried out under the Energy Policy and Conservation Act and the Energy Conservation and Production Act. It serves as a liaison between states and the Department of Energy on energy efficiency and renewable energy resource programs and encourages the transfer of findings relating to energy efficiency and renewable energy resource technologies. Meets twice a year.

Publications: Annual Report.

Technological Risks and Opportunities for Future Energy Supply and Demand Panel

Energy and Materials Program
Office of Technology Assessment
U.S. Congress
Washington, DC 20510-8025

Description: Established August 28, 1987, by the Office of Technology Assessment at the request of the U.S. House Committee on Energy and Commerce and its subcommittee on Fossil and Synthetic Fuels, and the U.S. House Committee on Government Operations. The panel has 17 members.

Purpose: Advisor to the Office of Technology Assessment.

Activities: Reviews and advises on plans, progress, and reports of the Office of Technology Assessment's study on the United States' future energy outlook, technology, and risks. The first part of the study deals with oil prices and the second part deals with supply and demand.

Forecasts for the years 1987 to 2010 are made. Holds two or three meetings a year.

Publications: A report.

Private Organizations in the United States

Alliance to Save Energy (ASE)
1725 K Street, NW, No. 914
Washington, DC 20006-1401

Description: Founded in 1977, ASE has a staff of 15 and 1,500 members.

Purpose: This is a group of business, government, and consumer leaders whose aim is to increase the efficiency of energy use.

Activities: Conduct research, have pilot projects and educational programs. They have an annual meeting.

Publications: Annual Report; also publishes reports, manuals, computer software, and other materials on energy topics.

American Association of Blacks in Energy (AABE)
801 Pennsylvania Avenue, SE, Suite 250
Washington, DC 20003

Description: Founded in 1977, the association has 500 members and a staff of one. There are six regional groups, 20 state groups, and 13 chapters. The budget is $170,000.

Purpose: Increase the knowledge, understanding, and awareness of the minority population in energy issues by serving as an energy information center for policymakers. The association is made up of blacks in energy-related professions, such as engineers, scientists, consultants, government officials, and academicians involved in energy use and research.

Activities: Encourage students to pursue professional careers in the field of energy, and the participation of blacks in energy programs and policy-making activities. The association keeps members informed regarding key legislation and regulations developed by the Department of Energy, the Department of the Interior, the Department of Commerce, and other federal and state agencies. A scholarship program is provided for higher education students. Information is provided on current jobs. The group maintains a speakers' bureau, operates archives, and bestows awards. There is a scholarship committee and a policy committee. Holds an annual conference.

Publications: Energy News (quarterly) and educational and research materials.

American Council for an Energy Efficient Economy (ACEEE)
1001 Connecticut Avenue, NW, Suite 535
Washington, DC 20036

Description: Founded in 1980, it is nonmembership with a $1 million budget.

Purpose: Evaluate and disseminate information to encourage use of energy efficient and economical technologies and practices.

Activities: Conducts research on energy conservation. Conferences are sponsored for the exchange of information among groups. Utilities, federal, state, and local energy officials, private industry, and consumers are given information on energy efficiency. The council holds a biennial summer study meeting on energy efficiency in buildings.

Publications: ACEEE Summer Study on Energy Efficiency in Buildings, Proceedings (biennial), *ACEEE Series on Energy Conservation and Energy Policy* (book series), and reports.

Americans for Energy Independence (AFEI)
1629 K Street, NW, Suite 602
Washington, DC 20006

Description: Founded in 1975, AFEI has a staff of three with a $220,000 budget. This is a group of business, labor, scientific, industrial, and academic individuals interested in energy independence.

Purpose: Works for U.S. independence in the field of energy.

Activities: Seeks a national energy policy with a defined set of objectives and priorities; cooperation between business, labor, and government; federal leadership in defining and achieving energy goals; gives energy issues high priority on the national policy agenda; and promotes reliance on domestic energy sources. Media tours are given and the "American Energy Update" (a radio program) is sponsored. Annual energy leadership awards are bestowed. AFEI conducts media seminars, Capitol Hill conferences, and educational activities to keep energy issues on the national agenda. Monthly conferences are held.

Publications: Energy Policy Series (periodic), *Newsletter* (quarterly), *Program Report* (annual), *The Electric Dilemma, Sunfuels, New Directions for Nuclear Power,* brochures, and position papers.

Association of Energy Engineers (AEE)
4025 Pleasantdale Road, Suite 420
Atlanta, GA 30340

Description: Founded in 1977, the Association has 7,000 members, a staff of nine, and 39 local groups. Members are engineers, architects, and professionals interested in energy management and cogeneration, and manufacturers and industries involved in energy.

Purpose: Promotes the development of the profession and also the development of professional members.

Activities: The association awards scholarships to outstanding students in energy engineering and conducts seminars. It sponsors the Cogeneration Institute, a division of 2,000 cogeneration professionals, and the Environmental Engineers and Managers Institute, a division with 800 members. The association absorbed the National Association of Professionals in Energy Conservation, Total Energy Management Professionals, and Energy Management and Controls Society. An annual meeting is held.

Publications: The Cogeneration Journal (periodic), *Energy Engineering Journal* (periodic), *Newsletter* (periodic), and *Strategic Planning for Energy and Environment.*

Association of Professional Energy Managers (APEM)

3104 O Street, Suite 301
Sacramento, CA 95816

Description: Founded in 1983, this association has 500 members and five regional groups.

Purpose: Promote sound energy management principles.

Activities: Professional seminars and special events are held to encourage the professional exchange of information and ideas that have an impact on the energy management profession. Holds an annual meeting.

Publications: APEM Membership Directory (annual), *Professional Energy Manager* (monthly), and a *Newsletter.*

Center for Energy Policy and Research (CEPR)

c/o New York Institute of Technology (NYT)
Old Westbury, NY 11568

Description: This nonmembership center was established by the New York Institute of Technology in 1975 and has a staff of five.

Purpose: Disseminate information and do research on energy utilization and conservation.

Activities: Conducts Master of Science in Energy Management and professional certificate programs through NYT's School of Management to provide special training in new approaches to energy conversion and utilization. Seminars are also conducted. No special meeting is held.

Publications: None.

Citizens Association for Sound Energy (CASE)
c/o Encyclopedia of Associations
Gale Research Inc.
P.O. Box 33477
Detroit, MI 48232-9852

Description: Founded in 1974, the association has 165 members who are interested in energy matters.

Purpose: Keep the public informed regarding energy matters.

Activities: Acts as intervenor in Dallas Power and Light rate hearings and operating license hearings for Comanche Peak nuclear power plant. The association has a library and speakers' bureau. No meeting is mentioned.

Publications: Newsletter (two–four/yr.), reprints, and pamphlets.

Cogeneration and Independent Power Coalition of America (CIPCA)
Two Lafayette Center
1133 21st Street, NW, Suite 420
Washington, DC 20036

Description: Founded in 1980, the coalition has 25 members and a staff of three. Members include industrial users of paper, steel, petroleum, and coke; energy development companies; gas utilities; electric utility subsidiaries; and engineering design firms.

Purpose: Promote financial and regulatory incentives to cogeneration and independent power project development.

Activities: Makes recommendations to the U.S. Congress and federal and state government departments and agencies. The coalition has a speaker's bureau. An annual meeting is held.

Publications: ACA/CIPCA Newsletter (periodic) and Conference *Proceedings* (periodic).

Energy Conservation Coalition (ECC)
1525 New Hampshire Avenue, NW
Washington, DC 20036

Description: Founded in 1981, the coalition has 20 members and 500 local groups.

Purpose: Promote the energy conservation.

Activities: Promotes the participation of residential, commercial, public utility, and industry in energy conservation. It focuses on promotion of favorable government policies, provision for development of energy-saving technologies, and removal of institutional barriers to conservation, such as building codes that hinder energy efficient building techniques.

Statistics and data reports on energy conservation issues are compiled. Bestows annual award for best conservation efforts by a legislator, individual, or a group. The coalition maintains a library. No meetings are listed.

Publications: Powerline (bimonthly).

Environmental Policy Institute (EPI)
218 D Street, SE
Washington, DC 20003

Description: Founded in 1974, the institute has a budget of $1.6 million. This is a division of Friends of the Earth with 14,000 members.

Purpose: Work for energy conservation and greater use of renewable sources of energy.

Activities: Acts as a national information center on energy and environmental issues. It sees that laws, once passed, are implemented by the Executive Branch. It keeps citizens informed about major developments in the Executive Branch and recommends public participation in the legislative process. The institute recommends conservation of energy and water resources. Periodic meetings are held.

Publications: Annual Report, EPI Quarterly Report, NGU Directory for International Environmental Activists (periodic), and publishes materials on environmental topics including energy, chemical safety, nuclear power and waste, agricultural resources, and water resources.

National Association of Energy Service Companies (NAESCO)
1440 New York Avenue, NW
Washington, DC 20005

Description: Founded in 1983, the association has 90 members. Members include those interested in the energy service industry.

Purpose: Promote the energy service industry.

Activities: Informs the public regarding performance contracting and financing of alternative energy and conservation programs. It promotes the development and growth of the energy service industry. It is an information clearing house and holds workshops. An annual meeting is held.

Publications: NAESCO News (bimonthly).

National Energy Management Institute (NEMI)
601 N. Fairfax Street, Suite 160
Alexandria, VA 22314

Description: Founded in 1981, this nonmembership institute has a staff of 40 and 11 state groups.

Purpose: Promote efficient use of energy.

Activities: Offers building owners and managers a warranty program that provides energy cost-savings projects that will make structures more efficient without diminishing any aesthetic or comfort variables. It analyzes the air quality of buildings. Some of the state groups sponsor seminars on energy. No mention of meetings is given.

Publications: News from NEMI (periodic).

National Energy Resources Organization (NERO)

11 Canal Center, Suite 250
Alexandria, VA 22314

Description: Founded in 1975, the organization has 450 members with a budget of $30,000.

Purpose: Promoting national energy activities.

Activities: The exchange and distribution of ideas on energy development, supply, and use. Recognition is made for people and organizations who have contributed to the field of energy production and alternative sources. Government and industry are encouraged to participate in advancement of energy technology. Annual awards are presented at the awards banquet. The organization sponsors monthly luncheons on Capitol Hill. Conferences/symposia are held annually.

Publications: None.

National Energy Specialist Association (NESA)

NESA Building
518 Gordon Street, NW
Topeka, KS 66608

Description: Founded in 1984, the association has 1,000 members and a staff of seven. Members are professionals involved in energy management, marketing, and sales.

Purpose: The association promotes technological advances to conserve energy to help foster U.S. energy independence.

Activities: Members are helped to increase sales and profits. Members exchange ideas on management, sales, and marketing and deal with consumer problems and complaints. It sponsors competitions and holds training programs. An annual meeting is held.

Publications: NESA Journal (quarterly) and a *Newsletter* (monthly).

Scientists and Engineers for Secure Energy (SE2)

570 Seventh Avenue, Suite 1007
New York, NY 10018

Description: Founded in 1976, the group has 1,200 members and a staff of one. Members include scientists, scholars, and engineers involved in the energy problem.

Purpose: Exchange information on energy issues.

Activities: The members' expertise on energy issues is offered to the public and governmental agencies responsible for the development of a legal framework for the solution to the issues. They provide the best possible insight to legislative, judicial, and regulatory processes. They sponsor educational forums on energy featuring distinguished scientists, interact with legislative and judicial branches of the government, and conduct scientific colloquia. No meeting is mentioned.

Publications: None.

U.S. Energy Association (USEA)
1620 I Street, Suite 615
Washington, DC 20006

Description: Founded in 1924, the association has 100 members and a staff of two. Members include those interested in energy in government, industry, educational institutions, and professional service organizations.

Purpose: Provide consideration of energy resources, policy, management, use, and conservation as related to the energy picture in United States and the world.

Activities: Publishes data on energy resources and their utilization. It holds conferences and forums for those concerned with surveying, developing, or using energy resources. Special energy seminars are held. The U.S. Energy Award is bestowed. USEA holds annual meetings and participates in the triennial World Energy Congress.

Publications: Annual Assessment of U.S. Energy Policy and Prospects, USEA Report (bimonthly), *U.S. Energy Association Annual Report,* and *National Energy Data Profiles.*

Western Interstate Energy Board/WINB (WIEB/WINB)
6500 Stapleton Plaza
3333 Quebec Street
Denver, CO 80207

Description: Founded in 1969, the board has 16 members and a staff of seven with a budget of $350,000. It includes the western states and has 16 state groups.

Purpose: Help western state governors and legislatures and the federal government with issues affecting energy resources in the West.

Activities: Fosters development of all energy resources in the West; developed a balance of performance and responsibility between the states and federal government. A library is maintained. It sponsors programs, conferences, regional studies, and organizes and implements cooperative projects. No meeting is scheduled.

Publications: Annual Report, Newsletter (biweekly), staff analyses, and special reports.

Women's Council on Energy and the Environment (WCEE)

P.O. Box 33211
Washington, DC 20033

Description: Founded in 1981, the council has 200 members and a budget of $10,000. Individuals who work for the government, industry, consulting firms, and the community educating the public about national policy issues are members.

Purpose: Educate the public about national policy issues concerning energy.

Activities: Promotes professional development of women interested in energy and environmental issues. It advocates decision-making on energy issues by business and government officials. The council maintains a speakers' bureau, sponsors monthly programs, roundtables, lectures, and panel discussions. It holds an annual meeting with seminars.

Publications: Membership Directory (annual) and *WCEE Newsletter* (monthly).

Women in Energy (WE)

c/o Kim Flesher
America Nuclear Society
555 N. Kensington
La Grange Park, IL 70525

Description: Founded in 1978, the association has 314 members, three regional groups, five state groups, and ten local groups with a budget less than $25,000. Members work in energy-related fields.

Purpose: Educate the public about complex energy choices and issues so decision-making is more responsible.

Activities: Organizes energy briefings and seminars, operates a speakers' bureau, sponsors exhibits, and arranges tours of energy facilities. WE also judges high school science fairs and awards certificates of recognition. An annual meeting is held.

Publications: Membership Directory (periodic), *Women in Energy News* (quarterly), educational publications, and audiovisuals.

Individual Energy Organizations

Fuel

Interagency Commission on Alternative Motor Fuels
Office of Deputy Secretary
Department of Energy
1000 Independence Avenue, SW
Washington, DC 20585

Description: Established by Public Law 100-494, the Alternative Motor Fuels Act of 1988. Membership includes the deputy secretary of energy, assistant secretary of defense, acting assistant secretary for occupational safety and health, assistant administrator for air and radiation, Commission of the Federal Supply Service, and the assistant postmaster general.

Purpose: Coordinate federal efforts to develop and put into use a national alternative motor fuels policy.

Activities: Ensure development of a long-term plan for commercialization of alcohols, natural gas, and other alternative motor fuels. The commission provides for the exchange of information among persons interested in alternative fuels.

Publications: None.

United States Alternative Motor Fuels Council
Interagency Commission on Alternative Motor Fuels
Office of the Deputy Secretary
Department of Energy
1000 Independence Avenue, SW
Washington, DC 20585

Description: Established by Public Law 100-494, the Alternative Motor Fuels Act of 1988. It has 20 members: two from the U.S. House, two from the Senate, and the other 16 represent state or local government bodies knowledgeable about alternative motor fuels.

Purpose: Serve as an advisor to the Interagency Commission on Alternative Motor Fuels.

Activities: Reports to and advises the Interagency Commission on Alternative Motor Fuels.

Publications: None.

Petroleum and Natural Gas

American Petroleum Institute (API)
1220 L Street, NW
Washington, DC 20005

Description: Founded in 1919, the institute has 200 members with a staff of 400. Corporations in the petroleum and allied products industries are members.

Purpose: Exchange information and foment cooperation between government and industry on all issues of national concern.

Activities: Promotes the petroleum industry, encourages study of arts and sciences dealing with the petroleum industry, conducts research on petroleum, maintains a large library, and operates a central abstracting and indexing service in New York City. An annual meeting is held.

Publications: API Directory (annual), Publications and Materials (annual), Report to the Membership (annual), *Basic Petroleum Data Book* (three updates/yr.), *Imported Crude Oil and Petroleum Products* (monthly), *Inventories of Natural Gas Liquids and Liquified Refinery Gases* (monthly), *Joint Association Survey* (annual), Monthly Completion Report, Monthly Statistical Report, Weekly Statistical Bulletin, Quarterly Completion Report, manuals, and booklets.

Energy Consumers and Producers Association (ECPA)
P.O. Box 1726
Seminole, OK 74868

Description: Founded in 1977, the association has 495 members and a staff of three. It includes small, independent producers and royalty owners.

Purpose: Works for proper treatment of independent producers and royalty owners.

Activities: Challenges court rulings and legislation set up by government agencies, works with congressional persons on exemptions for producers, and keeps congressional offices informed regarding the effects of legislation on the gas and oil industry. No meeting is scheduled.

Publications: Newsline (monthly).

Gas Research Institute (GRI)
8600 W. Bryn Mawr Avenue
Chicago, IL 60631

Description: Founded in 1976, the institute has 255 members and a staff of 269. Individuals involved with gas companies, distribution of gas, and interstate and intrastate lines are members.

Purpose: Keeps people informed about new developments in the field.

Activities: Research is done on natural and synthetic gas, distribution, safety, and environmental impacts. Workshops are held related to gas technology. An annual meeting is held and a biennial conference.

Publications: Baseline Projections of U.S. Natural Gas Supply and Demand (annual), *Gas Research Insights* (periodic), *Gas Research Institute* (annual report), *Gas Research Institute—Digest* (quarterly), *Gas Research Institute— Information Directory* (annual), *IGRC Proceedings* (periodic), and *Technology Profiles* (periodic).

Independent Petroleum Association of America (IPAA)
1101 16 Street, NW
Washington, DC 20036

Description: Founded in 1929, the association has 6,000 members and a staff of 25. Members are independent oil and gas operators and those independents interested in oil and gas.

Purpose: Encourage independent oil and gas producers.

Activities: Represents small oil and gas producers on the federal level, maintains a speakers' bureau, and compiles statistics. It holds semiannual meetings.

Publications: Executive Report (monthly), *The Oil and Natural Gas Production Industry in Your State* (annual), and *Petroleum Independent* (monthly).

Interstate Oil Compact Commission (IOCC)
P.O. Box 53127
Oklahoma City, OK 73152

Description: Founded in 1935, the commission has 700 members and a staff of eight with a $500,000 budget. The commission includes states that produce oil and gas.

Purpose: Dedicated to the conservation of energy resources.

Activities: Sets up rules and guidelines for the proper maintenance and production of wells, which saved millions of barrels of oil. There is a library of 2,000 volumes. Statistics are compiled. A semiannual conference is held.

Publications: Directory of Interstate Oil Commission and State Gas Agencies (periodic), *Interstate Oil and Gas Compact and Committee Bulletin* (periodic), *National Stripper Well Survey* (annual), and monograph series.

National Association of Royalty Owners (NARO)

119 N. Broadway
Ada, OK 74820

Description: Founded in 1980, the association has 5,000 members and a staff of five with a $200,000 budget. Members are owners of mineral properties and oil and gas royalties.

Purpose: Keep members informed about new legislation and tax information on mineral properties.

Activities: Assists owners in managing their mineral properties and gives information on tax and legislative matters. Mineral management seminars are held and a hall of fame is maintained. It compiles statistics and bestows awards. An annual meeting is held.

Publications: Annual Report, Hotline Alerts (periodic), *Royalty Owners Action Report* (monthly), *Tax Packets* (annual), monographs, and books.

Pennsylvania Grade Crude Oil Association (PGCOA)

c/o Pringle Powder Company
Box 201
Bradford, PA 16701

Description: Founded in 1923, the association has 200 members and a staff of one. Crude oil producers are members.

Purpose: Disseminate information regarding crude oil production.

Activities: Supports research on economical and efficient recovery of crude oil. An annual meeting is held.

Publications: None.

Petroleum Marketers Association of America (PMAA)

1120 Vermont Avenue, NW, Suite 1130
Washington, DC 20005

Description: Founded in 1941, the association has 43 members and a staff of 19. Wholesale marketers and retail fuel dealers are members.

Purpose: Promote favorable petroleum distribution competition among members.

Activities: Encourages the adequate supply of petroleum products to serve their customers. Management institutes for jobbers are held. An annual Distinguished Service Award is bestowed. An annual convention is held.

Publications: Journal of Petroleum Marketing (monthly), *Petroleum Marketing Databook* (semiannual), *PMMA Directory* (annual), *PMMA Weekly Review,* and surveys.

Society of Independent Gasoline Marketers of America (SIGMA)
Reston Town Center
11911 Freedom Drive, 1 Fountain Square
Reston, VA 22090-5103

Description: Founded in 1958, the society has 285 members and a budget of $1 million. Members are private wholesale and retail gasoline marketers.

Purpose: Disseminate information to the marketers.

Activities: Represents the interests of marketers before the government and legislative agencies. It provides statistical data on industry and conducts national meetings in marketing regions of the country. Semiannual meetings are held.

Publications: Independent Gasoline Marketing (bimonthly), *Society of Independent Gasoline Marketers—Roster* (annual), position papers, news releases, and information on industry issues.

Coal

American Coal Ash Association (ACAA)
1913 I Street, NW, 6th floor
Washington, DC 20001

Description: Founded in 1968, the association has 70 members and a staff of eight, with a $450,000 budget. Members include electric utility companies, engineering and research organizations, ash marketing companies, and coal transportation companies.

Purpose: Provide technical assistance to coal ash producers.

Activities: Does research on uses for ash, transportation of ash, and its disposal. The association advertises, does public relations programs, and disseminates information on ash production, use, and disposal. A 2,000-volume library is maintained. Statistics are compiled. Seminars are held. Biennial symposia are held.

Publications: Annual Report, Ash at Work (bimonthly), *Executive* (periodic), and *Process and Technical Data* (periodic).

American Coal Foundation (ACF)
1130 17th Street, NW, Suite 220
Washington, DC 10036

Description: Founded in 1981, the foundation has 49 members and a staff of two. Members are coal producers and suppliers, utility companies, and railroads.

Purpose: Promote the use of coal.

Activities: Provides information giving the advantages and uses of coal as a fuel. It provides classroom materials, sponsors workshops, and participates in coal conventions. No meeting is listed.

Publications: None.

American Coke and Coal Chemicals Institute (ACCCI)
1255 23rd Street, NW
Washington, DC 20037

Description: Founded in 1944, the institute has 65 members and a staff of four with a budget of $600,000. Members are interested in coke and coal products.

Purpose: Promote use of coke and coal products.

Activities: The institute has several active committees—chemicals, coke, foundry, tar products, traffic, and safety and health—whose functions are performed by members. An annual meeting is held.

Publications: Directory (periodic), *Foundry Facts* (three/yr.), and a quarterly *Newsletter.*

American Mining Congress (AMC)
1920 N Street, NW, Suite 300
Washington, DC 20036

Description: Founded in 1897, the congress has 400 members with a staff of four and a $5.5 million budget. Members are interested in the production of coal and the equipment used.

Purpose: Promote the mining industry.

Activities: Keeps producers, manufacturers of mining equipment, and institutions that serve the mining industry informed of new developments in the field. It bestows a Sentinels of Safety Award and a Distinguished Service Award. It operates a speakers' bureau and offers workshops and seminars. An annual meeting is held.

Publications: AMC Journal (monthly) and convention session papers.

Anthracite Industry Association (AIA)
805 15th Street, NW, Suite 310
Washington, DC 20005

Description: Founded in 1981, the association has 50 members. Members are miners and anyone who has any interest in anthracite.

Purpose: Promote the use of anthracite.

Activities: Makes people aware of the benefits of anthracite for heating purposes, acts as an advocate for the industry at state and federal levels, and works for expansion of the anthracite market by conducting market development campaigns. An annual meeting is held.

Publications: Newsbreaker (semiannual) and a *Newsletter.*

Association of Bituminous Contractors (ABC)
2020 K Street, Suite 800
Washington, DC 20006

Description: Founded in 1968, the association has 250 members. Members are independent and general contractors who build mines and mine facilities.

Purpose: Better the conditions for miners.

Activities: Facilitates collective bargaining with the International Union United Mine Workers of America. It promotes health and safety programs related to the coal mining industry. An annual meeting is held.

Publications: None.

BCR National Laboratory (BCRNL)
400 William Pitt Way
Pittsburgh, PA 15238

Description: Founded in 1934, the laboratory has a staff of 25.

Purpose: Disseminate information to coal-producing companies, chemical companies, and electric utilities.

Activities: Research in mining, preparation, transportation, energy issues, and the use of bituminous coal. Studies are done to find methods for control of air and water pollution due to coal mining. A library of 40,000 documents on coal is maintained.

Publications: Reports, pamphlets, and publications on coal-related topics.

Bituminous Coal Operators' Association (BCOA)
303 World Center Building
918 16th Street, NW
Washington, DC 20006

Description: Founded in 1916, the association has 16 members and a budget of $2 million. Members include firms engaged in mining bituminous coal.

Purpose: Promotes cooperation among coal operators.

Activities: Promotes industrial relations among operators and employees represented by the International Union United Mine Workers. It negotiates wages, hours, and conditions of employment. Mine safety and mining methods improvements are promoted. An annual meeting is held.

Publications: None.

Coal Exporters Association of the United States (CEA)
1120 17th Street, NW
Washington, DC 20036

Description: Founded in 1945, the association has 35 members. Members are exporters of coke and coal.

Purpose: Promote coal exportation.

Activities: A biennial conference is held.

Publications: Bulletin (biweekly).

Federal-State Coal Advisory Board
Bureau of Land Management
Department of the Interior
1849 C Street, NW
Washington, DC 20240

Description: Established March 20, 1980, by a charter signed by the secretary December 21, 1979. It consists of 23 members including the chairperson and governor (or designee) from Alabama, Colorado, Montana, New Mexico, Oklahoma, Utah, and Wyoming, and also the chairperson of each regional coal team.

Purpose: Promoting federal coal management.

Activities: Advises the secretary of the interior and the director of the Bureau of Land Management regarding the federal coal management program. The board meets once a year.

Publications: None.

Fort Union Regional Coal Team
Division of Mineral Resources
Bureau of Land Management
222 N. 32nd Street
P.O. Box 36800
Billings, MT 59107

Description: Established March 20, 1980, by a charter signed by the secretary of the interior December 21, 1979; rechartered in 1987 when it functioned independently of the Federal-State Coal Advisory Board. It consists of five voting and nonvoting members. Bureau of Land Management gives administrative support as staff.

Purpose: Oversight of the federal coal leasing program and advisory role to the Federal-State Coal Advisory Board.

Activities: Guides and oversees both competitive and noncompetitive coal leasing in the Fort Union region, gives the secretary alternative leasing levels in tons of recoverable coal to be offered for lease, guides preparation of the regional lease sale environment impact statement, considers land use plans to be used for regional coal activity planning, serves as a forum for the Department of the Interior and state consultation and cooperation in major coal management programs, and gives advice to the Federal-State Coal Advisory Board. The team meets at least once a year.

Publications: None.

Green River-Hams Fork Regional Coal Team
Branch of Solid Minerals
Bureau of Land Management
2850 Youngfield Street
Lakewood, CO 80215

Description: Established March 20, 1980, by a charter signed by the secretary of the interior December 21, 1979, and rechartered March 27, 1986, when it functioned independently of the Federal-State Coal Advisory Board. The team has five voting members, including the governors of Colorado and Wyoming, state directors of the Bureau of Land Management in Colorado and Wyoming, and the state director for Mineral Resources from Colorado's state office, Bureau of Land Management. One nonvoting member comes from the Washington, D.C., office and is appointed by the secretary of the interior. Administrative support is provided by the Bureau of Land Management.

Purpose: Keep the secretary of the interior informed about coal activity in the region and advise the Federal-State Coal Advisory Board.

Activities: Transmits alternative leasing levels in tons of recoverable coal to be offered for lease, guides the preparation of the regional lease sale environment impact statement, considers land use plans to be used for the region, serves as a forum for Department of the Interior and state consultation and cooperation in major coal management programs, and gives advice to the Federal-State Coal Advisory Board. Meets at least once a year.

Publications: None.

National Coal Association (NCA)
1130 17th Street, NW
Washington, DC 20036

Description: Founded in 1917, the association has 150 members and a staff of 52 with a budget of $6 million. Members include people interested in the coal industry.

Purpose: Promote the use of coal and keep members informed about new developments.

Activities: Serves as a liaison between the industry and federal government agencies, keeps industry informed about legislation and action taken regarding taxation, works to control air pollution, and collects and distributes statistics. An annual meeting is held.

Publications: Coal Bids (monthly), *Coal Data* (annual), *Coal Distribution Statistics* (annual), *Coal Industry Employment/Production* (annual); *Coal News* (weekly), *Coal Transportation Statistics* (annual), *Coal Voice* (bimonthly), *Electric Utility Coal Stockpiles* (annual), *International Coal* (annual), *International Coal Review* (monthly), *Steam Electric Market Analysis* (monthly), *Steam Electric Plant Factors* (annual), and educational materials.

National Coal Council (NCC)

2000 Fifteenth Street, N, Suite 500
Arlington, VA 22201

Description: Council was established October 25, 1984, by the secretary of energy. In 1984 it absorbed the functions of the Fossil Energy Advisory Committee. It has no more than 125 members appointed by the secretary of energy for two-year terms. Support services are provided by the Office of Fossil Energy.

Purpose: Serves as a public advisory council under the Department of Energy.

Activities: Provides recommendations to the secretary of energy on matters relating to coal, such as federal policies that affect production, marketing, and use of coal; plans, priorities, and strategies regarding technology relating to coal production and use; proper balance between various elements of federal coal-related programs; scientific and engineering aspects of coal technologies; and the progress of coal research and development, pursuant to Public Law 86-599. Subcommittees meet as needed. The council meets semiannually.

Publications: Improving International Competitiveness of U.S. Coal and Coal Technologies (1987), *Reserve Data Bank* (1987), *Coal Conversion* (1986), *Clean Coal Technology* (June 1986), *Interstate Transmission of Electricity* (June 1986), and *New Source Performance Standards on Industrial Boilers Report* (June 1986).

Pennsylvania Coal Association (PCA)

212 N. 3rd Street, Suite 102
Harrisburg, PA 17101

Description: Founded in 1988, the association has 220 members and a staff of eight. Coal companies, engineers, and equipment manufacturers are members.

Purpose: Informs members of new developments in the field.

Activities: Lobbies for the coal mining industry, informs members of state and federal laws and regulations, and bestows Reclamation Awards. An annual meeting is held.

Publications: Coal Data (annual), *Coal Quarterly,* and a *Directory* (annual).

Powder River Regional Coal Team

Wyoming Bureau of Land Management
Branch of Solid Minerals
2515 Warren Avenue
P.O. Box 828, MS925
Cheyenne, WY 82003-1828

Description: Established March 20, 1980, by a charter signed by the secretary of the interior December 21, 1979, it was rechartered May 6, 1988, when it became independent of the Federal-State Coal Advisory Board. The team has five voting members, and there are nonvoting members. Administrative support is provided by the Bureau of Land Management.

Purpose: Keep the secretary of the interior informed about coal activity in the region and advise the Federal-State Coal Advisory Board.

Activities: Transmits alternative leasing levels in tons of recoverable coal to be offered for lease, guides preparation of regional lease sale environment impact statement, considers land use plans to be used for regional coal activity planning, serves as a forum for the Department of the Interior and state consultation and cooperation in major coal management programs, and gives advice to the Federal-State Coal Advisory Board. Meets at least once a year.

Publications: None.

Rocky Mountain Coal Mining Institute (RMCMI)

3000 Youngfield, No. 324
Lakewood, CO 80215-6545

Description: Founded in 1912, the institute has 800 members and a staff of one with a $125,000 budget. Members include coal mine officials, state and federal inspectors, Bureau of Mines personnel, and anyone interested.

Purpose: Improve the mining industry.

Activities: Holds education programs pertaining to coal mining technologies and production in order to advance the mining industry. Scholarships are given to persons pursuing a mining career. The institute holds an annual meeting.

Publications: Proceedings of the RMCMI (annual).

San Juan River Regional Coal Team
Division of Mineral Resources
Bureau of Land Management
Rodeo Road Office
P.O. Box 1449
Santa Fe, NM 87504-1449

Description: Established May 20, 1980, by a charter signed by the secretary of the interior December 21, 1979, it was rechartered May 6, 1988, when it functioned independently of the Federal-State Coal Advisory Board. There are three members representing the Bureau of Land Management, one representing the state of New Mexico, one representing the state of Colorado, and 20 ex-officio members.

Purpose: Members are interested in coal activity of the region.

Activities: Oversees federal coal leasing activities in the San Juan River Region, including competitive and noncompetitive coal leasing. The team meets at least once a year.

Publications: None.

Uinta-Southwestern Utah Regional Coal Team
Bureau of Land Management
324 S. State Street, Suite 301
Salt Lake City, UT 84111-2303

Description: Established March 20, 1980, by a charter signed by the secretary of the interior December 21, 1979. It was inactive 1984 through June 4, 1987, when it was rechartered and now functions independently of the Federal-State Coal Advisory Board. There are ten members. Administrative support is provided by the Bureau of Land Management.

Purpose: Oversee coal activity in the region.

Activities: Oversees the federal coal leasing activities in the region, transmits alternative leasing levels in tons of recoverable coal to be offered for lease, guides preparation of regional lease sale environmental impact statements, considers land use plans to be used for regional coal activity planning, serves as a forum for the Department of the Interior and state consultation and cooperation in major coal management programs, and gives advice to the Federal-State Coal Advisory Board. Meets at least once a year.

Publications: None.

Western Coal Transportation Association (WCTA)
P.O. Box 176
Denver, CO 80201

Description: Founded in 1974, the association has 70 members. It includes electric utilities and coal-producing companies.

Purpose: Keep members informed about legislation and recent developments in the field.

Activities: Promotes the export of coal from the United States, facilitates the transportation of coal, and provides a forum for the exchange of ideas among members. Seminars are conducted on coal transport railroad cars. Studies are made on state and federal legislation affecting the coal industry. Scholarships and fellowships are awarded and research projects are sponsored. A library and speakers' bureau are maintained. Annual operation and maintenance seminars are held. An annual meeting is held.

Publications: General Meeting and Conference Transcript (annual), *Spring Meeting and Conference Transcript* (annual), brochures, and special studies.

Hydropower

Edison Electric Institute (EEI)
701 Pennsylvania Avenue, NW
Washington, DC 20004-2696

Description: Founded in 1933, the institute has 190 members and a staff of 300. Investor-owned electric utility companies in the United States are members.

Purpose: Promote the use and conservation of electricity.

Activities: Maintains a library and a speakers' bureau. It compiles statistics and sponsors educational programs. Awards are given. An annual meeting is held.

Publications: Electric Perspectives (bimonthly), *Electrical Reports* (weekly), *Rate Book* (annual), *Statistical Reports* (weekly), *Statistical Yearbook, Electric Power Surveys,* books, and pamphlets.

Electric Power Research Institute (EPRI)
3412 Hillview Avenue
Palo Alto, CA 94304

Description: Founded in 1972, the institute has 500 members and a staff of 700 with a budget of $259 million. Membership is composed of sectors of the electric utility industry.

Purpose: Develop an economically and environmentally acceptable program of research and development for electric power production.

Activities: Does research in the areas of coal combustion systems, advanced power systems, energy and environment, energy management and utilization, and nuclear power. It also maintains a library. No meeting is mentioned.

Publications: Guide (three/yr.) and *Journal* (nine/yr.)

National Hydropower Association (NHA)

555 13th Street, NW, Suite 900 East
Washington, DC 20004

Description: Founded in 1983, the association has 75 members, a staff of two, with a budget of $140,000. Members are those interested in hydropower development.

Purpose: Promote the development of hydroelectric power.

Activities: Makes the government aware of the potential of hydropower and helps eliminate barriers to its development. The association monitors and drafts new legislation and presents testimony to government regulatory and legislative bodies.

Publications: Hydro Regulatory Report (monthly), *NHA News from Washington* (monthly), a *Newsletter,* and position papers.

Nuclear Energy

American Nuclear Energy Council (ANEC)

410 First Street, SE
Washington, DC 20003

Description: Founded in 1975, the council has 125 members, a staff of 15, and a $3.5 million budget. Organizations interested in peaceful application of nuclear energy are members.

Purpose: Promote peaceful use of nuclear energy.

Activities: Supports the development of nuclear power as an energy source. It coordinates and projects the interests of the U.S. nuclear industry to Congress and the Executive Branch, and informs companies of actions taken by the Executive Branch and Congress on nuclear issues. An annual meeting is held.

Publications: None.

American Nuclear Society (ANS)

555 N. Kensington Avenue
La Grange Park, IL 60525

Description: Established 1954. Members: 14,000. Staff: 86. Local Groups: 51 (nine overseas). Student Branches: 53. Semiannual Meetings.

Purpose: To advance science and engineering in the nuclear industry, disseminate information, promote research, and establish scholarships.

Activities: Conducts meetings devoted to scientific and technical papers, works with governmental agencies, educational institutions, and other organizations dealing with nuclear issues. Maintains a library of 3,000 volumes on nuclear emergency and business management.

Committees include: Engineering and Technology Accreditation, Registration and Professional Development, Honors and Awards, International, Nuclear Engineering Education for the Disadvantaged, Professional Divisions, and Public Information.

Publications: ANS News (monthly), *Nuclear News* (monthly), *Nuclear Science and Engineering* (monthly), *Nuclear Standard News* (monthly), *Nuclear Technology* (monthly), *Fusion Technology* (bimonthly), *Transactions* (semiannual), and the ANS/DOE monograph series.

Americans for Nuclear Energy (AFNE)

2525 Wilson Boulevard
Arlington, VA 22201

Description: Founded in 1978, AFNE has 18,000 members, a staff of one, and a $200,000 budget. Members favor nuclear energy.

Purpose: Promote the use of nuclear energy.

Activities: Lobbies in Congress and state legislatures, rates Congress on nuclear issues, and pursues lawsuits for nuclear power. The group promotes prominent speakers and educates the public on the need for nuclear energy. Members hold periodic conferences.

Publications: Nuclear Advocate (bimonthly).

Citizen's Energy Council (CEC)

77 Homewood Avenue
Allendale, NH 07401

Description: Established 1966. Members: 2,200.

Purpose: Inform community leaders concerned with the hazards of nuclear energy.

Activities: Monitors and reports on activities of the Nuclear Regulatory Commission and other events related to nuclear energy. Conducts training sessions on community organization, organizes workshops. Has a 420-volume library on the history of the movement to oppose nuclear power.

Publications: Energy News Digest (newspaper, monthly) and *Last Chance Bulletin* (bimonthly).

European Nuclear Society (ENS)

(Europaische Kernenergie-Gesellschaft-ENS)
Postfach 2613
CH-3001
Berne, Switzerland

Description: Established 1975. Represents 17 national nuclear energy societies with 16,000 members, including European and non-European industrial plants, utility companies, and research institutions.

Purpose: Promote the interests of European nuclear energy producers and strive for scientific and engineering progress as it affects the peaceful use of nuclear energy.

Activities: Functions include keeping members informed about current developments in the nuclear energy industry worldwide and representing the industry before European and international bodies. Encourages cooperation among members. Provides scholarships, fellowships, and student exchanges. Conferences are also organized.

Publications: Nuclear Europe (periodic, English) and *Nuclear Newsletter for Euro-Politicians* (periodic)

Fusion Energy Foundation (FEF)

Address unknown

Description: Established 1974. Members: 20,000. Staff: 40.

Purpose: To provide a forum of independent, high-level scientific discussion about fusion from the standpoint of comprehensive policy-making.

Activities: Administers research programs and publishes articles concerned with historical aspects of the frontiers of science. Holds national and regional forums, seminars, and conferences. Provides speakers on nuclear and high technology development, economic development, and development of directed energy beam technologies for defensive weapons.

Publications: Fusion (bimonthly in English, French, German, Italian, and Spanish), *International Journal of Fusion Energy* (quarterly), and special reports.

International Atomic Energy Agency (IAEA) (Nuclear)

(Agence Internationale de l'Energie Atomique)
Vienna International Centre
Wagramerstrasse 5
Postfach 100
A-1400 Vienna, Austria

Description: Established: 1957. Membership: 113. Staff: 1,861.

Purpose: Formed to accelerate and enlarge the contributions of atomic energy to peace, health, and prosperity throughout the world. Ensures that any assistance given by the agency is not used for military applications.

Activities: Encourages research and development of the practical applications of atomic energy for peaceful uses throughout the world and fosters the exchange of scientific and technical information among nations. Has established health and safety standards and applies safeguards in accordance with the Treaty on the Nonproliferation of Nuclear Weapons.

Publications: Atomindex (semimonthly), *Nuclear Fusion* (monthly), *Bulletin* (quarterly), proceedings of conferences, symposiums, and seminars, technical directors, and safety and legal reports.

Renewable Energy

Alternative Sources of Energy (ASE)
107 S. Central Avenue
Milaca, MN 56353

Description: ASE was founded in 1971 and has a staff of ten.

Purpose: Publicize the practical application of alternative technologies for the independent power production industry.

Activities: It focuses on private power production technologies in cogeneration, biomass, waste-to-energy, geothermal, wind power, hydropower, and photovoltaics. The group has no meetings.

Publications: Independent Energy: The Business Magazine of the Independent Power Industry (ten/yr.)

American Solar Energy Society (ASES)
2400 Central Avenue, B-1
Boulder, CO 80301

Description: Founded in 1970, the membership is 2,000 with a staff of three. There are 26 regional groups. Members are those interested in solar energy.

Purpose: Promote the use and development of solar energy.

Activities: Does research, conducts workshops, holds discussions, and disseminates information to schools and the community. The library has 3,000 volumes. The society bestows awards. An annual meeting is held.

Publications: American Solar Energy Society—Membership Directory (annual), *Solar Today* (bimonthly), conference proceedings, and manuals for laypersons.

American Wind Energy Association (AWEA)
777 N. Capitol Street, NE, Suite 805
Washington, DC 20002

Description: Founded in 1974, the association has a staff of six with a $700,000 budget. The group is interested in wind energy.

Purpose: Promote the use of wind energy.

Activities: Promotes the use of energy from wind, wind energy to replace depletable fuels, and wind as a renewable source of energy. The association informs state and federal legislators about wind as a source of energy and does consulting work. Awards are bestowed. An annual meeting is held.

Publications: AWEA Wind Energy Weekly, Proceedings, *Windletter* (eight/yr.), *Wind Energy for a Growing World, Wind Energy in the Eighties: A Decade of Development,* and a brochure.

Biomass Energy Research Association (BERA)
1925 K Street, NW, Suite 503
Washington, DC 20006

Description: Founded in 1981, the association has 200 members. Members are corporations, people from universities, industries, and research laboratories with an interest in biomass and waste-to-energy fuels.

Purpose: Encourage biomass and waste-to-energy fuels research and commercialization.

Activities: Technology transfer, education in biomass energy research, and international cooperation in biomass research are promoted. A speakers' and consultants' bureau is maintained. The association holds bimonthly Capitol Hill luncheons on legislative policy issues and develops budget recommendations for congressional committees. An annual meeting with the Institute for Gas Technology is held along with periodic regional meetings.

Publications: Biologue (monthly) and special reports.

Conservation and Renewable Energy Inquiry and Referral Service (CAREIRS)
P.O. Box 8900
Silver Spring, MD 20907

Description: Founded in 1976, this nonmembership service has a staff of eight. It is a project of the U.S. Department of Energy.

Purpose: Act as an information clearing house.

Activities: Answers inquiries from the public, disseminates information about the use of renewable energy technologies and conservation for residential and commercial needs, assists the public in determining the feasibility of these technologies, disseminates information on solar heating, photovoltaics, biomass conversion, and ocean and wind energy. A network of public and private organizations that specialize in technical information is maintained. The service operates a 1,000-volume library. No meetings are held.

Publications: Distributes fact sheets, bibliographies, and brochures.

Energy Research Institute (ERI)
6850 Rattlesnake Hammock Road
Highway 951
Naples, FL 33962

Description: Founded in 1980, the institute has 3,900 members and a staff of four with a budget less than $25,000. Membership includes those interested in alternative energy sources.

Purpose: To achieve energy independence through development of projects dealing with wind generators, solar cells and collectors, steam generators, and hydrogen production with special focus on alcohol as a fuel.

Activities: Does research in alcohol production and crop management. Seminars and on-site experiments promote their research. Awards are given, children's services and a Hall of Fame are maintained. Statistics are compiled and they give guidance for local alcofuel club organizers. The institute holds annual meetings.

Publications: Directory (periodic) and reports (semimonthly).

Fusion Power Associates (FPA)
Two Professional Drive, Suite 248
Gaithersburg, MD 20879

Description: Founded in 1979, this association has 400 members and a staff of three. Members include those interested in fusion energy.

Purpose: Promote the development of fusion power as an energy option in the future.

Activities: Cooperation in fusion research among public and private organizations is promoted to develop an understanding of the potential of fusion energy. Members present testimony at congressional and appropriation hearings on fusion budgets. Research programs to assist in the technical progress of fusion development are held. Leadership awards are presented annually and certificates of appreciation are given to those who made major contributions to the development of fusion energy. Technical and managerial symposia for persons involved with fusion research are held. They have an annual meeting.

Publications: Fusion Facilities Directory (biennial), *Fusion Power Associates Proceedings* (annual), *Fusion Power Associates Executive Newsletter* (monthly), *What's News in Fusion* (quarterly), *Prospects for Fusion Power* (book), and audiovisual information.

Interstate Solar Coordination Council (ISCC)

900 American Center Building
St. Paul, MN 55101

Description: Founded in 1979, the council has 28 members including managers of state government renewable energy programs and others interested in the field.

Purpose: Exchange ideas and information regarding renewable energy programs.

Activities: Establishes a set of rules and regulations for industry programs and evaluates new systems, promotes cooperation between regulatory agencies and those affected by their decisions, develops use of solar cells by state government offices, and serves as a clearinghouse. The council holds annual meetings.

Publications: None.

National Wood Energy Association (NWEA)

777 N. Capitol Street, NE Suite 805
Washington, DC 20002

Description: Founded in 1979, the association has 300 members and a staff of three with a $120,000 budget. Public and private corporations interested in wood resources are included.

Purpose: Develop wood as a renewable energy source.

Activities: Draws consumers' and legislators' attention to wood as the greatest untapped renewable energy resource in the United States. It urges good forest, agricultural, and land management practices to develop wood as a renewable energy. The association conducts wood energy seminars and bestows awards. An annual meeting is held.

Publications: Biofuels Directory (periodic), *Biologue: Regional Biomass Program Reports* (five/yr.), *NWEA National Biomass Facilities Directory, Federal Programs for Research and Development of Biomass and Municipal Waste Technology.*

Natural Power

5420 Mayfield Road, Suite 205
Cleveland, OH 44124

Description: Founded in 1975, the group has 400 members, one regional group, one state group, and one local group. The members include those interested in using natural energy forces such as the sun, wind, and tides.

Purpose: Scientists, engineers, developers, and environmental activists want to replace toxic fuels and nuclear power with natural energy forces—sun, wind, tides, and currents.

Activities: Serves as an information exchange center and supports research. Public officials are influenced by letters and petitions. A speakers' bureau is maintained. Educational packets are provided for elementary and high school students. Awards are given. An annual meeting is held.

Publications: Natural POWER Newsletter (monthly) and articles.

Renewable Fuels Association (RFA)
201 Massachusetts Avenue, NE, C4
Washington, DC 20002

Description: Founded in 1981, the association has 50 members and a staff of four. There are 40 state groups and seven local groups. Members are interested in renewable fuels.

Purpose: Promote development of renewable fuels.

Activities: Works toward development in biomass fuel technology, mainly alcohol fuels. The association represents the renewable fuels industry before the government. It is active in the production of biomass fuel from agricultural crops. An annual meeting is held.

Publications: Newsletter (monthly).

International Organizations

Circum-Pacific Council for Energy and Mineral Resources (CPCEMR)
Halbouty Center
5100 Westheimer Road, Suite 500
Houston, TX 77056

Description: Founded in 1972, the council has a staff of one. It is multinational and includes organizations and governmental agencies of countries bordering the Pacific Ocean.

Purpose: Disseminate information on energy and minerals in the area.

Activities: Supports cooperative efforts in the study of energy and mineral resources in the region, has independent programs on the resources of the area, and strengthens communication among members. It holds periodic symposiums. An annual meeting is held.

Publications: None.

International Association for Energy Economics (IAEE)
1101 14th Street, NW, Suite 1100
Washington, DC 20005

Description: Founded in 1977, the association has 2,200 members and a budget of $150,000. Members are interested in energy economics.

Purpose: Individuals in government, universities, research organizations and energy industries promote energy economics.

Activities: Provides a forum for communication and exchange of experience and ideas in the field. Maintains Energy Economics Educational Foundation. Awards are presented to those who made outstanding contributions and wrote many pieces about energy economics. Semiannual meetings are held.

Publications: Energy Journal (quarterly), *Membership Directory* (biennial), *Newsletter* (quarterly), proceedings.

International Association for Hydrogen Energy (IAHE)
P.O. Box 248266
Coral Gables, FL 33124

Description: Founded in 1974, the association has 2,500 members and a staff of 12. Scientists, engineers, and professional societies involved in the use of hydrogen are members.

Purpose: Promote the use of hydrogen.

Activities: Promotes discussion and publication of ideas to further a clean energy system based on hydrogen. Independent and joint research projects are organized. They offer short courses and seminars on hydrogen energy. IAHE holds biennial meetings.

Publications: International Journal of Hydrogen Energy (monthly), *Miami International Conference on Alternative Energy Sources* (monthly), *Hydrogen Energy/Progress,* and proceedings of seminars and conferences.

International Association of Geophysical Contractors (IAGC)
P.O. Box 46209
Houston, TX 77056

Description: Founded in 1971, the association has 176 members and a budget of $240,000. The members are interested in petroleum exploration.

Purpose: Develop geophysical petroleum exploration in 20 countries.

Activities: Fosters the development of geophysics in the petroleum exploration industry and establishes guidelines for efficiency, growth, and safety. Research programs are conducted and seminars are held.

Publications: Newsletter (quarterly), *Special Report to Members* (periodic), manuals, brochures, and books.

International Atomic Energy Agency (IAEA)
Vienna International Centre
Wagramerstrasse 5
Postfach 100
A-1400 Vienna, Austria

Description: Founded in 1957, the agency has 112 members and a staff of 2,171 with a budget of $157,540,000. It is multinational.

Purpose: Ensure the peaceful use of atomic energy.

Activities: Encourages research on the peaceful uses of atomic energy, exchanges scientific information among nations, and operates laboratories and a research institute. It maintains a library in conjunction with organizations of the United Nations comprising 78,916 books, 592,806 technical reports, 4,187 periodicals, film titles, and documents.

Publications: Atomindex (semimonthly), *Bulletin* (quarterly), *CINDA* (semi-annual), *Meetings on Atomic Energy* (quarterly), *Newsbriefs* (monthly), *Nuclear Fusion* (monthly), proceedings of conferences, symposia, and seminars, technical directories, and safety legal and technical reports series.

International Federation of Petroleum and Chemical Workers (IFPCW)
P.O. Box 6565
Denver, CO 80206

Description: Founded in 1954, the federation is multinational and has a staff of one. It includes national unions representing 2 million workers in the petroleum and chemical industries.

Purpose: Promote cooperation among members.

Activities: Maintains a library. It is currently undergoing reorganization. An annual meeting is held.

Publications: None.

International Oil Scouts Association (IOSA)
P.O. Box 272949
Houston, TX 77277

Description: Founded in 1974, the association has 500 members and a staff of five with a $35,000 budget. It is a multinational association and includes 13 U.S. district associations of oil scouts.

Purpose: Disseminate information on well exploration.

Activities: Statistics are compiled on exploratory and development wells in the United States and the Free World, and on production in the oil and gas fields. Professional development and scholarship programs are offered. The association maintains a collection of yearbooks. An annual meeting is held.

Publications: International Oil and Gas Development Yearbook, International Oil Scouts Association—Annual Meeting Magazine, Newsletter (quarterly), *Oil Scouts Directory* (annual).

International Solar Energy Society (ISES)

P.O. Box 124
Caulfield East
VIC 3145 Australia

Description: Founded in 1954, the society has 4,100 members with a staff of three. There are 27 national groups. Members are interested in solar energy.

Purpose: Promote solar energy.

Activities: Affiliated with American Solar Energy Society. The principal activity is publication of various materials. It bestows awards. A biennial conference is held.

Publications: Proceedings of the International Congress (biennial), *ISES News* (quarterly), *Solar Energy Journal* (monthly), and *Sunworld* (quarterly).

Organization of the Petroleum Exporting Countries (OPEC)

Obere-Donaustrasse 93
A-1020 Vienna, Austria

Description: Founded in 1960, OPEC has 13 members and a budget of 234,906,844 Austrian shillings. It is multinational.

Purpose: Coordination and unification of the petroleum policies of member countries to protect their interests.

Activities: Works toward stabilization of prices in international oil markets to eliminate fluctuations while keeping the interests of producing nations in mind, a steady income for producing countries, a regular supply to consumer nations, and a fair return on capital for those investing in the petroleum industry. OPEC maintains a library, conducts research, and compiles statistics. Semiannual meetings are held.

Publications: Bulletin (monthly), *Review* (quarterly), *Statistical Bulletin* (annual), books, and booklets.

Organization Sources

Encyclopedia of Associations, Vol. 1, National Organizations of the U.S. Detroit, MI: Gale Research, 1991.

Encyclopedia of Associations, Vol. 4. International Associations. Detroit, MI: Gale Research, 1991.

Encyclopedia of Governmental Advisory Organizations, 1992-1993. 8th ed. Detroit, MI: Gale Research, 1991.

The United States Government Manual, 1991/92. Washington, DC: U.S. Government Printing Office, 1991.

5

Bibliography

THERE IS A WIDE SPECTRUM OF ENERGY SOURCES. The traditional fossil fuels—coal, petroleum, and natural gas—have dominated the energy world in the past and still do today. Atomic energy was introduced in the 1950s and has become a major source of energy. The use of traditional fuels and, more recently, nuclear energy present potential environmental problems. Renewable energy sources remain in an elementary stage of development. As a response to varying levels in energy consumption, the literature presents a wide range of viewpoints. This bibliography presents a broad perspective in the modern development of energy resources.

Reference Works

Reference

Gowen, Marcia M., and Herb Wade. **Renewable Energy Assessments: An Energy Planner's Manual.** Honolulu, HI: Pacific Islands Development Program/Resource Systems Institute, East-West Center, 1985, 227 p. ISBN 0-86638-065-5.

This book was written as a quick reference on major elements considered in renewable energy planning. Estimating renewable energy potential is very difficult. Good energy planning is essential to development and economic growth. Resource supply and technology feasibility must be assessed. The manual has two sections. The first gives a background on energy and the economic definitions and methods used in making energy assessments. The second section discusses fuel and technology assessments. Many figures and tables are used in the book. Also given is a list of abbreviations, acronyms, and currency conversions. Two appendices and a short bibliography are included.

Hunt, V. Daniel. **The Gasohol Handbook.** New York: Industrial Press, 1981, 580 p. ISBN 0-8311-1137-2.

This handbook begins by explaining what gasohol is and the decision to produce it. It then considers the basic feedstock needed for production and the production processes and plant designs. Since ethanol is a flammable product, a special chapter is devoted to environmental and safety impacts. The handbook finally considers the availability of markets, feedstock cost and availability, water power and laws, and legislation. The book has reference sources and a bibliography. There is a glossary and list of acronyms and several tables give technical reference data.

Thumann, Albert. **Handbook of Energy Engineering**. 2nd ed. Lilburn, GA: The Fairmont Press, 1991, 429 p. ISBN 0-88173-124-2.

This excellent handbook begins by explaining energy engineering. Scientific knowledge must be used for the improvement of the overall use of energy. An energy engineer must have engineering skills in electrical, mechanical, and process engineering and knowledge in management. A short list of references is given. The book is illustrated with many figures and many tables.

Bibliographies

Cheremisinoff, Nicholas P., and Paul N. Cheremisinoff, eds. **Gasohol Sourcebook.** Ann Arbor, MI: Ann Arbor Science Publishers, 1981, 221 p. ISBN 0-250-40425-7.

This volume is a literary survey of published materials in biomass-alcohol developments. It is divided into five parts: biotechnology and bioconversion; ethanol and methanol production; automotive and other fuels; production of chemical feedstocks; and the economics of alcohol production. References go as far back as 1965 with many being abstracted. There is an author index.

Evans, John. **OPEC: Its Member States and the World Energy Market.** A Keesing's Reference Publication. Detroit, MI: Gale Research, 1986, 679 p. ISBN 0-582-90267-3.

OPEC is a very familiar, but not very well understood, international organization. This volume examines the structure of supply and demand in the world market economy, the commercial structure of the world oil industry, the political and economic situation of each OPEC country, and the political and strategic issues arising from the importance of the world's principal source of energy. This book is a good reference source, giving the history of OPEC. It is divided into cross-referenced sections for easy use. It has many tables, maps, and diagrams to help in its use. A list of abbreviations and acronyms for oil companies as well as a general list of abbreviations is included.

Hosni, Djehana A. **Manpower for Energy Production: An International Guide to Sources with Annotations.** Bibliographies and Indexes in Economics and Economic History, No. 5, Westport, CT: Greenwood Press, 1986, 154 p. ISBN 0-313-25089-8.

This bibliography contains many references dealing with the United States, stressing the energy-producing sector that includes exploration, transportation, processing for oil and gas, and solar thermal industries. Energy-related activities, including research and development and supply and demand, are dealt with. There are 280 entries that include books, journals, government articles, and reports by research firms and universities. The coverage extends from the oil crisis to the present time, incorporating the oil price plunge and the Chernobyl accident. An appendix gives the country classification. A directory of publishers is also included.

Mandel, Rolfe D., and M. Elizabeth Hines. **Reclamation Planning for Coal-Mined Lands: A Selective Bibliography.** CPL Bibliography No. 151. Chicago: Council of Planning Librarians, 1985, 41 p. ISBN 0-86602-151-5.

This bibliography has 343 entries on coal and is divided into six parts: general information, regional perspectives for land reclamation, revegetating mined lands, reclamation and land use, reclaiming land for fish and wildlife, and innovative reclamation techniques. Entries include books, journal articles, and technical reports and are not annotated.

McCarl, Henry N., and David McConnell. **Bibliography on Energy Economics, 1975–1985.** Public Administration Series Bibliography No. P1876. Monticello, IL: Vance Bibliographies, 1986, 114 p. ISBN 0-89028-766-X.

This bibliography covers the period 1975–1985 and includes technical journals, books, and government publications that deal with energy economics. The bibliography is arranged alphabetically by author. It should be of interest to researchers, public officials, and anyone interested in the world energy situation.

Rohwedder, W. J. **Energy-Efficient Planning: An Annotated Bibliography.** CPL Bibliography No. 141. Chicago: Council of Planning Librarians, 1984, 29 p. ISBN 0-86602-141-8.

This small bibliography is a collection of selected resources in the area of energy planning. Material includes books, journal articles, and handbooks. The bibliography is divided into six main parts: community and bioregional planning; land use, transportation, and zoning; site and building design; landscape design; Third World applications; and bibliographies and directories. Each entry is annotated.

Vance, Mary. **Refuse as Fuel: A Bibliography.** Public Administration Series. Bibliography No. P2846. Monticello, IL: Vance Bibliographies, 1990, 20 p. ISBN 0-7920-0506-2.

Vance has compiled an alphabetical list of books, journal articles, and government documents about using waste products to produce energy.

White, Anthony G. **Economic Impacts of Energy Shortfalls: A Selected Bibliography.** Public Administration Series Bibliography No. P2847. Monticello, IL: Vance Bibliographies, 1990, 9 p. ISBN 0-7920-0507-4.

White has compiled a short bibliography of books, magazine articles, and government documents about energy shortfalls, particularly among the electric power systems. Outages occur during a physical event such as storms and earthquakes. Residential systems fail if overloaded. In planning, load growth and demand must be considered.

Books

Energy—General

Ambler, Marjane. **Breaking the Iron Bonds: Indian Control of Energy Development.** Development of Western Resources. Lawrence, KS: University Press of Kansas, 1990, 351 p. ISBN 0-7006-0422-7.

Ambler received a fellowship from the Alicia Patterson Foundation to enable her to spend a year traveling to western energy reservations. She also received other financial assistance for research. Her book focuses on the 15 western reservations that have the most coal, uranium, oil, and gas (excluding natives of Alaska). She begins the book with the history of federal Indian policies, economic development, and reservation energy development. She then focuses on the efforts tribes and allottees made during the 1970s and 1980s to increase control over energy development by using their power as mineral owners, as governments, and as partners in development. They do not oppose energy development. At the beginning there is a map showing the energy tribe reservations with coal, oil and gas, and uranium sites. Several pictures are found in the book and four appendices. Many pages of notes and a selected bibliography of books, articles, and government documents as well as personal interviews are listed.

Brinkman, Norman D., E. Eugene Ecklund, and Roberta J. Nichols, eds. **Fuel Methanol: A Decade of Progress.** PT-36. Warrendale, PA: Society of Automotive Engineers, 1990, 415 p. ISBN 1-56091-011-9.

This volume includes selected papers through 1989 on methanol as a fuel. Methanol is a promising alternative fuel because it has the potential to improve air quality, enhance the energy security of nations, and reduce ozone levels in urban areas. The volume includes studies done by experts from government, the energy industry, automotive industry, manufacturers, and universities. The book has an extensive bibliography of related readings and serves as a good guide for automotive industry professionals.

Brown, Lester R., Christopher Flavin, and Colin Norman. **The Future of the Automobile in the Oil-Short World.** Worldwatch Paper, 32, Washington, DC: Worldwatch Institute, 1979, 64 p. ISBN 0-916468-31-3.

This small volume shows how the changing world oil outlook affects the design and role of the automobile. Government is funding money for construction of highways and public transportation systems. Office buildings are now providing parking garages. It shows how the depletion of the earth's oil reserves is changing the outlook for the automobile. The book does have a good bibliography. There are a few tables used as illustrations.

Byrne, John, and Daniel Rich, eds. **The Politics of Energy Research and Development.** Energy Policy Studies, Vol. 3. New Brunswick, NJ: Transaction Books, 1986, 181 p. ISBN 0-88738-653-9.

Several authors, each an expert in his area, have contributed to this volume. The first chapter discusses the tragedy of U.S. energy R&D policy which was formed without regard to national energy needs The second chapter examines the origin of the nuclear power fiasco. Another chapter discusses the peaceful use of nuclear explosives. One chapter is devoted to solar energy and another to the role played by the government in conservation research and development. The R&D policy orientation in the United States got its inspiration from the military R&D programs of World War II, especially the Manhattan Project. This book should be read to understand the tragedy of our failure to relieve our dependence on imported oil. Each chapter has bibliographic references. A few tables are found in the book.

Calzonetti, Frank J., and Barry D. Solomon. **Geographical Dimensions of Energy.** The GeoJournal Library. Boston, MA: D. Reidel Publishing, 1985, 516 p. ISBN 90-277-2061-4.

The Energy Specialty Group of the Association of American Geographers contributed to this manuscript about energy geography, which investigates the spatial implications of energy. The book is divided into six parts. Part I examines thematic literature reviews on primary energy resources. Part II discusses economic, behavioral, and planning issues for production of major energy resources. Part III deals with the siting of power plants. Part IV focuses on energy consumption issues. Part V is concerned with multiregional and environmental issues. Part VI focuses on research areas that need to be studied and the future of energy geography. References are found with each chapter.

Devine, Michael D. **Energy from the West: A Technology Assessment of Western Energy Resource Development.** Science and Public Policy Program. Norman, OK: University of Oklahoma Press, 1981, 362 p. ISBN 0-8061-1750-8.

This book is based on the results of "Technology Assessment of Western Energy Resource Development," a study done for the U.S. Environmental Protection Agency by a research team from the Science and Public Policy Program at the University of Oklahoma. The study identified and analyzed consequences of the development of energy resources in the western United States and evaluated ways to deal with the consequences. The book has photos, tables, and figures. A list of bibliographic references is found at the end of each chapter.

Frisch, J. R. **Energy 2000–2020: World Prospects and Regional Stresses.** Translated by P. Ruttley. London, England: Graham and Trotman, 1983, 259 p. ISBN 0-86010-482-6.

The Conservation Commission of the World Energy Conference met in New Delhi in 1989 to study the world energy problem. More than 50 experts were involved in the study, which divided the world into ten regions with the United States being one. It was their hope that coal would be a good substitute for oil as the most used energy. However, nuclear power, natural gas, and hydropower were studied as substitutes for oil. New energies do emerge, so they studied the possibility of wood, vegetable, and animal waste. The delegates made a recommendation to the International Executive Council in each of the regions to maintain a small, permanent study group to continue the study of specific problems in greater depth. The book has a very small bibliography. There are 25 annexes.

Gonzalez, Richard J., Raymond W. Smilor, and Joel Darmstadter, eds. **Improving U.S. Energy Security.** Cambridge, MA: Ballinger Publishing, 1985, 308 p. ISBN 0-88730-015-4.

This volume was compiled from contributions by several authors. The first part examines where we've been, where we are, and where we're going. The second discusses the problem of increasing supply and reducing demand. Part three shows the impact of technology, 1984–2000, and the final part considers initiatives that should be examined. The book has many figures and tables and a list of abbreviations and acronyms. Most of the chapters have bibliographic notes.

Hall, Charles A. S., Cutler J. Cleveland, and Robert Kaufmann. **Energy and Resource Quality: The Ecology of the Economic Process.** Environmental Science and Technology, a Wiley-Interscience Publication. New York: Wiley, 1986, 577 p. ISBN 0-471-08790-4.

This book is clearly written to help readers understand the physical meaning of resource scarcity and the impact it will have. Part one gives a general discussion of energy, part two discusses energy and economics, part three shows the characteristics and magnitude of U.S. energy resource systems; part four examines human and environmental impacts of energy extraction and use; part five discusses energy and the management of renewable natural resources. The book is well illustrated with figures and tables. There is an extensive bibliography.

Helm, John L., ed. **Energy, Production, Consumption, and Consequences.** Washington, DC: National Academy Press, 1990, 296 p. ISBN 0-309-04077-9.

This book is based on a symposium, "An Energy Agenda for the 1990s," the first of a series of symposia to celebrate the opening of the Arnold and Mabel Beckman Center of the National Academies of Science and Engineering in Irvine, California, sponsored by the Program Offiice of the National Academy of Engineering. Contributors to the volume analyzed energy systems from the perspective of supply and demand, environmental effects, and the evolving vulnerabilities and opportunities of the oil industry, natural gas, and nuclear energy, and the growing role of electricity. Many figures and tables enhance the book. References are included with each chapter. A short biographical sketch of each contributor is included.

Kapstein, Ethan B. **The Insecure Alliance: Energy Crises and Western Politics Since 1944.** New York: Oxford University Press, 1990, 257 p. ISBN 0-19-50581-8.

This book tells the story of energy shortages and the importance of alliance relationships. The author then shows why alliance energy security has proved to be an elusive goal. The book has notes for each chapter and a selected bibliography. Several tables and figures are found throughout the book.

Knoepfel, Heinz. **Energy 2000: An Overview of the World's Energy Resources in the Decades to Come.** New York: Gordon and Breach Science Publishers, 1986, 181 p. ISBN 2-88124-074-7.

Energy is an essential force for our mere existence. Energy is necessary for food, housing, transportation, industry, and agriculture. There are five chapters in the book discussing energy today, energy in mankind's evolution, nuclear fission energy, energy tomorrow, and the future of energy. The future solutions to our energy problems should not be based solely on cost but on human, ecological, and ethical considerations. The book has many figures and tables, and a list of references for each chapter.

Knowles, Ruth Sheldon. **America's Energy Famine: Its Cause and Cure.** Norman, OK: University of Oklahoma Press, 1980, 351 p. No ISBN.

Knowles wrote this book in an easy reading style telling how we became a great nation because of our energy development and use, what our technical achievements and political failures are, how we became dependent on foreign oil, and what we need to do to achieve energy self-sufficiency. She did a great deal of research to write this book, collecting material from books, journals, congressional hearings, and newspapers. Some of these items are listed in the preface. She sees the ability to achieve an energy supply position where the United States is not vulnerable to embargoes or price increases as a realistic goal. Scientists agree that the United States still has great resources to be found onshore and offshore. Solar energy and wind power energy will be developed.

Kraushaar, Jack J., and Robert A. Ristinen. **Energy and Problems of a Technical Society.** Rev. ed. New York: John Wiley & Sons, 1988, 512 p. ISBN 0-471-61409-2.

This book is the result of a course taught by the authors at the University of Colorado, Boulder. It is intended for nonscience people and aims to bring forth an understanding of the technological problems of our society. In the book the authors discuss such topics as energy fundamentals, energy from fossil fuels, electric power, nuclear energy and safety aspects, uses of solar energy, alternative sources of energy, conservation, transportation of energy, and other energy topics. The book is illustrated with many tables and figures. References and problems are found at the end of each chapter. Answers to the problems are found in an appendix.

Lee, Thomas H., Ben C. Ball, Jr., and Richard D. Tabors. **Energy Aftermath.** Boston, MA: Harvard Business School Press, 1990, 274 p. ISBN 0-87584-219-4.

The authors discuss the history of the U.S. energy sector over the last forty years. They investigated the mistakes that were made and how they can be prevented in the future. Part one delves into changes in the energy situation from 1945–1989 and the government and corporate blunders made. Part two tells what went wrong and the lessons learned. Part three discusses the structure of energy systems over the next few decades. The book has many figures and tables. Notes and references are found at the end of each chapter.

Maull, Hanns W. **Energy, Minerals and Western Security.** Baltimore, MD: Johns Hopkins University Press, 1984, 413 p. ISBN 0-8018-2500-8.

The book begins by discussing the part minerals play in the economic security of a country. The lengthy energy chapter dwells on uranium, solid fuels, natural gas, oil, and the risks encountered. The author also discusses nonfuel minerals such as iron ore, titanium, cobalt, manganese, aluminum, and others. The last chapter deals with the economic security risks in minerals and strategies to deal with these risks and threats. A glossary gives both technical and nontechnical terms. There are tables and figures for each chapter. The bibliography is in the form of notes at the end of each chapter. The book has both a subject and author index.

Munasinghe, Mohan, and Gunter Schramm. **Energy Economics, Demand Management and Conservation Policy.** New York: Van Nostrand Reinhold, 1983, 464 p. ISBN 0-442-25838-0.

The authors wrote this book for people involved in energy work, such as planners, economists, policymakers, teachers, and researchers. The first part of the book presents background information and an analytical base, while the second part deals with specific case studies representing countries of various sizes, geographic locations, climates, energy resource endowments, and usage patterns. Some figures and tables are found throughout the book. There are a few footnotes, but the book lacks a bibliography.

Peirce, William Spangar. **Economics of the Energy Industries.** Belmont, CA: Wadsworth Publishing, 1986, 265 p. ISBN 0-534-05286-X.

This volume gives an overview of the changes that took place in energy markets since the crisis of the 1970s. The reserves and resources of the extractive industries (coal, oil, and natural gas) are discussed, followed by a discussion of other sources of energy such as nuclear power, electricity, solar, and biomass. The final section examines the problems encountered and the part played by the government in solving the problems. The book has bibliographic notes at the end of each chapter. Many tables and figures are found throughout the book.

Sant, Roger W., Dennis W. Bakke, Roger F. Naill, with James Bishop, Jr., eds. **Creating Abundance: America's Least-Cost Energy Strategy.** New York: McGraw-Hill, 1984, 176 p. ISBN 0-07-041518-8.

Sant and Bakke created the Mellon Institute's Energy Productivity Center in 1977 to develop and test different ways to treat energy as any other commodity whose price and supply fluctuate. Naill became a coauthor about a year later. The nation's energy system is different from the national defense system or the space program where the government assumes operating responsibility, while energy is produced and consumed mostly in the private sector. The authors feel the energy problems are being solved. They show that conservation is a "source" of energy and that free-market forces and least-cost strategies can make it happen. The book is illustrated with many figures. Many bibliographic footnotes are used throughout the book. A list of conversion factors can be found at the beginning to help the reader.

Treat, John E., ed. **Energy Futures: Trading Opportunities for the 1990s.** Tulsa, OK: Penn Well Publishing, 1990, 466 p. ISBN 0-87814-347-5.

This book is a compilation of chapters by many authors. The authors examined such aspects as futures, past and present; natural gas futures; principles of technical analysis; oil market fundamentals; options; risk management in the energy industry; an overview of governmental regulations over future markets; effect that energy futures will have on world oil markets; and where energy futures are headed. The book has eight appendices and a glossary.

Waters, L. L. **Energy to Move.** Owensboro, KY: Texas Gas Transmission Corporation, 1985, 240 p. No ISBN.

This book tells the story of Texas Gas Transmission Corporation from its origin to its merger with CSX Corporation. The author got his information for the book from such sources as employee magazines, annual reports, and personal interviews with executives of the firm. Tables are given in the appendices. Many pictures showing the firm are found in the middle of the book.

Energy Policy

Deacon, Robert T., and others. **Taxing Energy: Oil Severance Taxation and the Economy.** Independent Studies in Political Economy. New York: Holmes & Meier, 1990, 161 p. ISBN 0-8419-1179-7.

This book is written in a nontechnical style to be understood by the general public. All mathematical and analytical material appears in the appendices. Interstate comparisons of tax and government royalty burdens on crude oil production are discussed along with the issue of tax exporting. A chapter analyzes the relationship between the effects of tax and price changes on crude-oil supply. An estimation of the employment and government revenue effects of the tax is examined. Many figures and tables are used to prove the authors' views. The book has bibliographic references.

Devine, Michael D., and others. **Cogeneration and Decentralized Electricity Production: Technology, Economics and Policy.** Westview Special Studies in Natural Resources and Energy Management. Boulder, CO: Westview Press, 1987, 303 p. ISBN 0-8133-7287-9.

The National Science Foundation supported the study necessary to produce this book. It gives the history of the electric utility industry and tells why decentralized electricity production is important. Also discussed is the federal and state role in decentralized electricity production. There are excellent chapters on cogeneration, waste and biomass, small-scale hydropower, and wind power. The book has bibliographic notes and references as well as figures and tables.

Ender, Richard L., and John Choon Kim, eds. **Energy Resources Development: Politics and Policies.** New York: Quorum Books, 1987, 232 p. ISBN 0-89930-216-5.

Many authors contributed to this volume, which covers the diverse set of energy issues from both policy and political perspectives. This book was written after the crisis of the 1970s and before the fall of prices in the mid-1980s. Energy is not only vital to our economy and national security, but has a definite relationship with domestic policies such as transportation, housing, environment, social policies, and foreign trade. Congress had power over 60 percent of U.S. energy supplies, as oil and gas produced domestically provided 30 percent of U.S. needs, so Congress was faced with the problem of whether to continue controls or allow prices to rise. Conservation became an important issue. Many tables and figures are used throughout the book. References are found with each chapter.

Energy Imperatives for the 1990s. Washington, DC: The Atlantic Council of the United States, 1990, 53 p. No ISBN.

This small volume is the Report of the Atlantic Council's Energy Working Group. This project was funded by the Amoco Foundation, Bechtel Corporation, Chevron USA, Inc., ERC International, Phillips Petroleum, Ruhrgas A.G., Shell USA, Westinghouse, and the U.S. Department of Energy. Energy policy issues necessary for economic growth and environmental protection are examined and many recommendations were made to carry out the policies. Its theme is that: "The United States can utilize energy in environmentally responsible ways to support sustainable non-inflationary economic growth." The publication has four appendices.

Gray, John E., Henry H. Fowler, and Joseph W. Harned. **U.S. Energy Policy and U.S. Foreign Policy in the 1980s.** Cambridge, MA: Ballinger Publishing, 1981, 311 p. ISBN 0-88410-902-X.

This is a report of the Atlantic Council's Energy Policy Committee that was composed of 53 members. They studied the energy problem—oil. Oil is an economic, national security, and shared problem. A study was made of the issues associated with U.S. international relationships on energy and recommended actions to be taken by the United States and its allies. They studied United States' energy relations with Canada, Mexico, Venezuela, Japan, and Arab producers. The book has a glossary, appendix, and list of members.

Gray, John E., and Yoshihiro Nakayama, eds. **U.S.-Japan Energy Policy Considerations for the 1990s.** Lanham, MD: University Press of America, 1988, 104 p. ISBN 0-8191-7095-X.

This book is the result of a conference held at the Department of State, Washington, D.C., November 11–13, 1987. It was a joint project of the Atlantic Council of the United States, the Japanese Committee for Energy Policy Promotion, and the Japanese Institute of Energy Economics. U.S.-Japan energy relationships, gas, electric power, oil, coal, nuclear power, and solar energy were discussed. The book lists both United States and Japanese members of the joint working group.

Guertin, Donald L., W. Kenneth Davis, and John E. Gray, eds. **U.S. Energy Imperatives for the 1990s: Leadership, Efficiency, Environmental Responsibility, and Sustained Economic Growth.** A Book of the Atlantic Council of the United States. Lanham, MD: The Atlantic Council of the United States and University Press of America, 1992, 280 p. ISBN 0-8191-8336-9.

This book forms a part of the Atlantic Council's overall program on energy and the environment. Background is provided on such critical issues as energy-related environmental issues, energy efficiency, the supply of liquified fuel-oil, gas and alternative fuels, electric power demand and supply, electric power generation and nuclear power, and the need for the United States to be a leader in the energy area.

Kash, Don E., and Robert W. Rycroft. **U.S. Energy Policy: Crisis and Complacency.** Norman, OK: University of Oklahoma Press, 1984, 334 p. ISBN 0-8061-1869-5.

This book was written to examine the nation's recent energy history and the attempt to formulate a national energy policy as a result of the energy crisis. The U.S. policy process failed to respond effectively to the crisis. The four primary sources of energy (oil, natural gas, coal, and nuclear energy) are surveyed, covering the period 1973–1980. The book then discusses alternatives such as hydroelectric power, oil shale, geothermal energy, electricity, and solar energy. The book ends by giving several recommendations that would help to provide a secure energy future. The book has references for each chapter. Figures are used as illustrations.

Kemezis, Paul, and Ernest J. Wilson, III. **The Decade of Energy Policy: Policy Analysis in Oil-Importing Countries.** Praeger Scientific. New York: Praeger Publishers, 1984, 271 p. ISBN 0-03-062783-4.

The authors show the radical changes in the international oil market between 1973 and 1983 and how governments in oil-importing countries adapted their domestic energy policies to meet the changes. They then stressed the role that oil imports play in the process of adaptation. Three chapters discuss the government's efforts to have greater control over both the international and domestic energy supplies. Demand management policies pursued by importing countries are examined. Finally, the politics and administration of energy policy between 1973 and 1983 are analyzed. There is a bibliography including books, articles, documents, and unpublished materials.

Lowinger, Thomas C. **Energy Policy in an Era of Limits.** Praeger Special Studies. Praeger Scientific. New York: Praeger Publishers, 1983, 223 p. ISBN 0-03-060423-0.

The purpose of this book is to evaluate the formulation and implementation of U.S. energy policy in the post oil embargo period and why the United States failed to have an effective set of energy policies in the past decade. An effective energy policy should reinforce the signs provided by the free energy markets rather than turn off their impacts. Two chapters discuss the effects of government intervention in the petroleum and natural gas markets. Another chapter examines the state of the nuclear industry and its future. One chapter presents an overview of energy conservation policies in the United States. The final chapter relates ways to have a more effective energy policy. The book has a few tables and figures, and bibliographic references.

Resources for the Future. **U.S. Energy Policies: An Agenda for Research.** Baltimore, MD: Johns Hopkins University Press, 1968, 152 p. No ISBN.

This staff report begins by giving a general picture of the energy industries. The demand for energy will increase and the future supply of the various sources of energy is unpredictable. Separate chapters include oil, natural gas, coal, electricity, nuclear energy, and shale oil. The final chapter deals with the interrelationships among the differing energy sources. Several tables are used in the book. Bibliographic items are missing.

Sawyer, Stephen W., and John R. Armstrong, eds. **State Energy Policy: Current Issues, Future Directions.** Westview Special Studies in Natural

Resources and Energy Management. Boulder, CO: Westview Press, 1985, 307 p. ISBN 0-8133-7027-2.

Several authors contributed to this book. The authors review the current issues in state energy policy and discuss the directions that will be taken in the future. Environmental impacts of energy extraction and utilizatio-have been reduced, but fossil fuel combustion continues as the major source of air pollutants. States have become active in trying to solve the energy problem to protect the environment. Energy production, consumption, and expenditures are phenomena that occur within all states. Energy is a multibillion dollar issue in most states. Most states have gone beyond the federal policy to develop their own policies and programs dealing with energy. Bibliographic references are found at the end of each chapter. The book has many tables and figures.

Shaw, T. L., D. E. Lennard, and P. M. S. Jones, eds. **Policy and Development of Energy Resources.** A Wiley-Interscience Publication. New York: Wiley, 1984, 247 p. ISBN 0-471-10537-6.

Many authors, experts in specific areas of energy, contributed to the compilation of this volume. This book was published after the end of the cheap oil era. It discusses such topics as where to place the emphasis on research and development, how demands are affected by technological advances, the influence of political and financial institutions on the development and exploitation of new resources, understanding the availability of such resources as coal, oil, gas, nuclear power industries, and renewable energy sources, and the storage and conservation of energy. Figures and tables are interspersed throughout the book. Each chapter has several bibliographic references.

Stobaugh, Robert B., and Daniel Yergin, eds. **Energy Future.** 3rd ed. New York: Vintage Books, 1981, 459 p. ISBN 0-394-74750-X.

This volume is a report of the Energy Project conducted at the Harvard Business School. It was first published in 1979 and has been revised for this third edition. In researching the book, business executives, labor union leaders, analysts and government officials were consulted. Many changes occurred in the energy field since the first edition was published. Such changes as the upheaval in the Middle East, the revolution in Iran and the war between Iran and Iraq, Mexico's entry as an oil producer, and a falling demand for OPEC oil, among others, occurred. In this volume the authors tried to integrate the changes with the many things that did not change in order to understand the fundamental questions involving the energy future. The goal was to provide a framework for thinking about the future of energy. Coal, natural gas, nuclear power, solar energy, and conservation are discussed in length. Many pages of references are found for the chapters.

Tugwell, Franklin. **The Energy Crisis and the American Political Economy: Politics and Markets in the Management of Natural Resources.** Stanford, CA: Stanford University Press, 1988, 294 p. ISBN 0-8047-1500-9.

The author worked on many energy problems in the Department of State, the Congressional Office of Technology Assessment, and the U.S. Agency for International Development from 1973 to 1983. He pulls together what he learned from these experiences, trading the historical roots of our actions and the government arrangements that shaped them while investigating possible solutions. The book is divided into four parts: a background in which he gives the history of the oil, coal, natural gas, and electricity industries; crisis and response during the Nixon, Ford, Carter, and Reagan administrations, and the Arab oil embargo; assessment where he discusses how we did as a country manage the energy crisis; and, finally, he gives his conclusions. Several pages of notes, followed by a lengthy bibliography, complement the text.

Vietor, Richard H. K. **Energy Policy in America Since 1945: A Study of Business-Government Relations.** Studies in Economic History and Policy. New York: Cambridge University Press, 1984, 363 p. ISBN 0-521-26658-0.

Vietor examines the fossil-fuel energy policy in the United States since World War II. He discusses the transition to peace and fluid fuels, 1945–1958; the problems of managing surplus through political consideration 1959–1968; and the second energy transition—adjustment to depletion, 1969–1980. The book has many charts, tables, and figures. Numerous footnotes serve as bibliographic items.

Conservation

Barkovich, Barbara R. **Regulatory Interventionism in the Utility Industry: Fairness, Efficiency, and the Pursuit of Energy Conservation.** New York: Quorum Books, 1989, 181 p. ISBN 0-89930-383-8.

The book begins with a review of the history of energy/utility regulation in the United States, emphasizing California regulations. It then examines theories of regulation and factors that influence regulatory decision-making. It presents a case study of regulatory intervention through the strategy devised by the California Public Utilities Commission to promote energy conservation in California's investor-owned energy utilities.

Major utility conservation programs in California—residential weather-izing financing, commercial incentives and audits, residential appliance incentives, and conservation voltage regulation—are discussed fully. A short bibliography and bibliographic notes are found at the end of each chapter. The book has a few illustrative tables.

Chandler, William U. **Energy Productivity: Key to Environmental Protection and Economic Progress.** Worldwatch Paper, 63. Washington, DC: Worldwatch Institute, 1985, 62 p. ISBN 0-916468-63-1.

This small volume discusses the upheavals in world energy markets in the 1970s that caused a conservation revolution. It takes many years to supply new sources of energy, so energy must not be wasted. Economic progress and development are dependent upon energy productivity. Many bibliographic notes are found at the end of the text.

Darmstadter, Joel, Hans H. Landsberg, Herbert C. Morton, and Michael J. Coda. **Energy Today and Tomorrow: Living with Uncertainty.** Englewood Cliffs, NJ: Prentice-Hall, 1983, 233 p. ISBN 0-13-277640-5.

The authors begin by telling why energy has become one of society's major concerns and the ways in which we use energy. Energy resources and new technology applied to make resources more adequate are described. A major chapter is devoted to the competition in energy industries and the pros and cons of government regulation. The impact energy activities have on our environment and the part energy plays in an unstable world are discussed. This book has many tables, figures, and plates, a conversion table, and glossary. Chapters have bibliographic notes and suggestions for further reading.

Gellings, Clark W., and Dilip R. Limaye, eds. **Electric Utility Conservation Programs.** New York: Praeger Publishers, 1986, 255 p. ISBN 0-275-92243-X.

Several authors contributed to this book. Many utilities have adopted strategic conservation programs to increase the efficiency of energy utilization to benefit both the utility and the customer. Because of the complexity and costs of conservation programs, electric utilities need to share information gained in the planning, implementation, marketing, and evaluation of conservation programs. This book disseminates information on conservation options for electric utilities. Utilities need to restore better customer relationships since they face the problem of improved energy efficient devices, conversion to natural gas and other fuels. The book is well illustrated with tables and figures. Bibliographic notes or references are found with each chapter.

Vine, Edward, and Drury Crawley, eds. **State of the Art of Energy Efficiency: Future Directions.** Series on Energy Conservation and Energy Policy. Washington, DC, and Berkeley, CA: American Council for an Energy Efficient Economy, 1991, 287 p. ISBN 0-918249-11-2.

The authors of chapters in this book discuss plans for future research and programs to promote energy efficiency in buildings. Such topics as design, implementation of government and utility programs, appliance standards, collecting and analyzing energy data, and resource planning are discussed. Various types of residential and commercial programs offered by electric utilities are examined. Technology involved in monitoring building energy performance is discussed. Information found in this book is of special value to professionals in utilities, government, research institutions, and universities and may be used to help them develop their own ideas for improving energy efficiency in buildings. Chapters have bibliographies. The book is illustrated with tables and figures.

Electrical Energy

Baughman, Martin L., Paul L. Joskow, and Dilip P. Kamat. **Electric Power in the United States: Models and Policy Analysis.** The MIT Press Energy Laboratory Series, 2. Cambridge, MA: MIT Press, 1979, 266 p. ISBN 0-262-02130-7.

Research for this book was funded primarily from National Science Foundation grants. It shows the need for a computer-based simulation model of electric power in the United States to show how public policies affect the supply and demand of electricity. The book should be of interest to people in industry, government, and education who are involved in the development of energy simulation models. Numerous policy simulations shed light on issues of fuel utilization, environmental policy, the future of nuclear energy and the price of electricity. The book has tables, figures and appendices. Bibliographical references and notes are given for each chapter.

Bolet, Adela M., ed. **Forecasting U.S. Electricity Demand: Trends and Methodologies.** CSIS Energy Policy Series. Boulder, CO: Westview Press, 1985, 273 p. ISBN 0-8133-7036-1.

The authors analyze the challenges and opportunities that confront the electric power industry by reconciling the differences among forecasters as to the future of electricity demand in the residential, commercial, and industrial areas. The book is divided into three parts based on electricity markets: industrial, residential, and commercial. Each section has several chapters by various authors presenting their forecasts for the use of electricity. The book is well illustrated with tables and figures. Bibliographic notes are found at the end of each chapter.

Calzonetti, Frank J., and others. **Power from the Appalachians.** Contributions in Economics and Economic History, No. 89. Westport, CT: Greenwood Press, 1989, 234 p. ISBN 0-313-25797-3.

This study, which considered electricity trade from the Appalachians to the Northeast, was done for the Energy and Water Research Center by eight faculty members from five disciplines (economics, electrical engineering, geography, mineral resources economics, and political science) at West Virginia University. The study assessed the role of electricity exports on the economy of West Virginia and whether the electricity exports should be increased. This book considers bulk power purchases focusing on the northeastern and midwestern regions of the United States. Nuclear powered facilities in highly populated regions and concern over acid rain are considered. Power exports from the coal fields can be increased either by higher utilization of existing capacity or by constructing new generating plants. The authors explain their outlook on electricity trade in the region as a means to promote regional solutions to regional problems of the northeast. A bibliography is found at the end of each chapter. Many tables and figures are used for explanatory purposes. The book has eight appendices for further explanation.

Flavin, Christopher. **Electricity for a Developing World: New Directions.** Worldwatch Paper, 70. Washington, DC: Worldwatch Institute, 1986, 66 p. ISBN 0-916468-71-2.

Flavin shows the importance of electric power in industrial countries. He discusses the development of electric power, the efficient use of electricity, new approaches to rural electrification, and decentralizing generators and institutions. The book is illustrated with a few tables. The bibliography includes almost one hundred items.

————. **Electricity's Future: The Shift to Efficiency and Small-Scale Power.** Worldwatch Paper, 61. Washington, DC: Worldwatch Institute, 1984, 70 p. ISBN 0-916468-61-5.

Electric power is a controversial energy issue. Nuclear energy as an electricity source has suffered setbacks. Coal-powered power plants are faced with problems of air pollution. Even so, electricity's energy use has risen during the past decade, replacing oil in some cases. A few utilities have encouraged small-scale power production, especially in California.

Large industrial companies are building their own power systems and tapping new energy sources such as wind power and geothermal energy. Small-scale power sources, including cogeneration, geothermal energy and wind power will supply most of the new generating capacity in California. This small volume has over one hundred bibliographic items.

————. **Electricity from Sunlight: The Future of Photovoltaics.** Worldwatch Paper, 52. Washington, DC: Worldwatch Institute, 1982, 63 p. ISBN 0-916468-50-X.

Flavin discusses the importance of solar photovoltaics as an expanding source of energy. Photovoltaics is a proven technology with a record of reliability, but had a major challenge when this book was written; the cost. Oil companies see this as an emerging industry in which they can invest some of their capital. The book has many bibliographic notes.

Gellings, Clark W., and Dilip R. Limaye, eds. **Strategic Marketing for Electric Utilities.** Lilburn, GA: Fairmont Press, 1988, 451 p. ISBN 0-88173-037-8.

The purpose of this book is to familiarize the utility practitioner with methods and techniques used to: implement marketing strategies, understand the customer's needs and preferences, understand the customer's decision-making process, understand the mechanics of establishing successful marketing programs; and, understand the challenges faced by electric utilities in converting from conservation orientation to the marketing focus needed in the next decade. Case studies of utility marking programs are given along with chapters to stimulate the reader so he will give some thought to the future directions of utility marketing. The book has many tables and figures. References are given for each chapter. The editors are contributors along with many authors.

National Research Council. Committee on Electricity in Economic Growth, Energy Engineering Board. **Electricity in Economic Growth.** Washington, DC: National Academy Press, 1986, 165 p. ISBN 0-309-03677-1.

The study was prepared by a group of experts serving on the Department of Energy's Committee on Electricity in Economic Growth so they could better understand the relationship between economic activity and electricity use. The purpose of this study is to show how changes in the economy can affect the growth of electric demands. The second objective is to examine the connection between the use of electrotechnologies and productivity. The publication has tables, figures, and references. It also has four appendices and a glossary.

Technology Futures Inc., and Scientific Foresight Inc. **Principles for Electric Power Policy.** Westport, CT: Quorum Books, 1984, 448 p. ISBN 0-89930-095-2.

Electric power is a major element of life today and will continue to be even greater in the future. Government policy will control the role it plays. Prior to the mid-1970s, electric power policy makers worked in a stable environment but, since then, the national rate of electricity demand growth has dropped. Nuclear generating facilities and coal-based power generation have come under attack. In mid-1981 the National Science Foundation funded a two-year, multidisciplinary technology assessment of alternative U.S. electric power futures and helped shape the nation's electric power policy. The book has tables and figures and a bibliography is found at the end of each chapter. Four appendices give views of expert panelists.

Wyatt, Alan. **Electric Power: Challenges and Choices.** Toronto, Ontario: Book Press, 1986, 186 p. ISBN 0-920650-00-7.

Energy is used by man for many purposes—for heating and cooling, for light, for industry, for transportation, and many other uses. Man's great use of energy has been made possible by his mastery of electricity, which can be transmitted over long distances and distributed economically to remote corners. Electricity is a convenient form of energy. This book, written in an easy reading style, should be of interest to utility stock and bond holders, employees of the industry, and anyone interested in electric power. The book is well illustrated with photos and tables. It contains a glossary, conversion factors, and abbreviations used in the text. A short list of recommended reading is given.

Petroleum and Natural Gas Industry

Banks, Ferdinand E. **The Political Economy of Natural Gas.** Croom Helm Commodity Series. New York: Croom Helm, 1987, 200 p. ISBN 0-7099-3940-X.

This is a book on applied energy economics, examining natural gas markets. The book gives a picture of the general energy outlook and also considers some macroeconomic aspects of the energy crisis. Natural gas has been called the "Prince of Hydrocarbons" as well as "Gas Fatal"; at one time it was responsible for many deaths each year. The book discusses gas markets in Europe, OPEC countries, and the Pacific. It also has an excellent chapter about natural gas in the United States. Many figures, tables, and bibliographic references are found in the book.

Bromley, Simon. **American Hegemony and World Oil: The Industry, the State System and the World Economy.** University Park, PA: The Pennsylvania State University Press, 1991, 316 p. ISBN 0-271-00746-X.

This volume is a socialists approval through a case study relating to world oil and the postwar hegemony of the United States. Emphasis is placed on the dramatic events of the 1970s that transformed the world oil industry, the subsequent and apparently related economic crisis, and the escalation of U.S. military expenditures during the renewal of superpower confrontation in the Second Cold War. As a consequence, the longevity and durability of the United States's global leadership is questioned. The study develops such themes as United States hegemony, oil as a strategic commodity, British and United States leadership in the Middle East, the economic crises of the 1970s and the role of OPEC, petropolitics in Europe, Japan, and the Soviet Union, and the U.S. strategy in the 1980s. This book challenges the neoclassical theories on which these accounts of world politics are based. The arguments presented contest both the theoretical basis of this evolution and also the substantial judgments that evolved. Extensive bibliographic notes are given for each chapter. There are several tables in the book.

Bull-Berg, Hans Jacob. **American International Oil Policy: Casual Factors and Effect.** New York: St. Martins Press, 1987, 209 p. ISBN 0-86187-696-2.

This book evolved over a period of five years and develops an insight into the political mechanics of American international oil policy during the period of high priced crude oil from 1973 to early 1986. The book is organized into two parts: Part I, "Explaining United States Foreign Oil Policies," and Part II, "The American Oil System Strategy." Each part can be used independently. There are a few tables and a list of abbreviations used in the book. The bibliography contains books, articles, and reports.

Chester, Edward W. **United States Oil Policy and Diplomacy.** Contributions in Economics and Economic History, No. 52. Westport, CT: Greenwood Press, 1983, 399 p. ISBN 0-313-23174-5.

Chester gives an overview of the historical evolution of U.S. oil policy. Some feel the multinational companies have played an exploitative role and the oil grants exercise considerable power over foreign nations, as well as American oil policy and diplomacy. U.S. petroleum firms were disturbed that foreign oil companies were favored over American ones. Petroleum-producing countries place their own national interest first and Americans play the role of a client searching for oil. There is an interesting checklist of name changes and abbreviations of petroleum companies. The chronology contains key events pertaining to petroleum. Bibliographical notes are given for each chapter followed by a bibliographic essay.

Clark, John G. **Energy and the Federal Government: Fossil Fuel Policies, 1900-1946.** Urbana, IL: University of Illinois Press, 1987, 511 p. ISBN 0-252-01295-X.

Clark begins by showing the competition between the coal, natural gas, and oil industries and the government policies toward the fuel industries during the World War I period. With the end of the war, strike after strike closed many coal mines causing shortages and price increases. During the 1920s' energy transition, the federal role continued in both the coal and oil industries. Clark then addresses the Depression years (1929–1930s) and discusses the federal policies adopted to try to stabilize fossil fuel industries. The last part of the book deals with the part the government played in the fossil fuels market and the many agencies established to administer policies during World War II. This book has several tables, figures, charts, and maps. Bibliographic notes are found for each chapter.

Ghanem, Shukri M. **OPEC: The Rise and Fall of an Exclusive Club.** New York: Methuen, 1986, 233 p. ISBN 0-7103-0175-8.

The modern oil industry began when Colonel Edwin Drake proved that oil existed beneath the earth's crust and discovered the first underground oil near Titusville, Pennsylvania, in 1859. From that time forward, oil companies emerged rapidly, leading up to the 1960 creation of the Organization of Petroleum Exporting Countries. The main reason for the creation of OPEC was to put an end to downward price fluctuations. In the 1980s there was a decline of OPEC power. The book has references for each chapter and many tables are presented.

Ghosh, Arabinda. **OPEC, the Petroleum Industry, and the United States Energy Policy.** Westport, CT: Quorum Books, 1983, 206 p. ISBN 0-89930-010-3.

This interesting volume tells the story of OPEC. It begins with OPEC's formation, pricing mechanisms, and its relationship with multinational oil companies. The U.S. oil industry, with its changing structure and performance in relation to OPEC, is discussed. Excellent chapters deal with Henry Kissinger and the oil crisis and how Nixon and Ford reacted to the energy crisis. Another chapter discusses the energy policy of Carter and Reagan. The final chapter gives the future of OPEC. The book has figures, tables, an abbreviations list, and a selected bibliography.

Ikenberry, G. John. **Reasons of State: Oil Politics and the Capacities of American Government.** Cornell Studies in Political Economy. Ithaca, NY: Cornell University Press, 1988, 213 p. ISBN 0-8014-2155-1.

Price shocks generated both national and international problems. Policy responses to the oil shocks in the United States were shaped by the institutional structures of the state. The opportunities and constraints that institutional structures mold for state action influence the ways in which state officials seek to solve adjustment problems. The United States turns to market pricing to achieve energy adjustment, which shows the importance of the market as a tool of state capacity. The extension or maintenance of markets can be a powerful tool of the state. The role and dynamics of institutional change and how institutional structures are shaped must be understood. The state is an institutional structure that gives government officials a unique standing in the pursuit of their own goals. The book is well footnoted and has a few illustrations.

Jentleson, Bruce W. **Pipeline Politics: The Complex Political Economy of East-West Energy Trade.** Cornell Studies in Political Economy. Ithaca, NY: Cornell University Press, 1986, 263 p. ISBN 0-8014-1923-9.

This book examines the issue of whether the United States and its NATO allies should trade with the Soviet Union. The controversy over the Siberian natural gas pipeline made the problem more complex. Chapter 1 sets up the framework to analyze East-West energy trade. Chapter 2 examines the Cold War period. Chapter 3 discusses the 1950s East-West trade growth period. Chapter 4 analyzes the 1962–1963 oil trade sanctions while Chapter 5 discusses the development of East-West trade in oil and natural gas in the 1970s. Chapter 6 discusses Reagan's sanctions against the Siberian natural gas pipeline. The last chapter gives some thoughts on the East-West trade issue. An abbreviations list can be found at the beginning of the book as well as a list of tables. Bibliographic items are also recorded.

Kalter, Robert J., and William A. Vogely, eds. **Energy Supply and Government Policy.** Ithaca, NY: Cornell University Press, 1976, 356 p. ISBN 0-8014-0966-7.

Fossil fuels (oil, natural gas, and coal) provide the primary energy supplies of the United States. Other supplies come from atomic power and hydropower. In order to increase the supply of energy exploration, development, and production, refining and transportation must be better developed. The government employs six types of policies to control private market relationships. They are taxation, federal lands, research and development, environmental policies, price policy, and trade policy. Many authors, specialists in their fields, contributed to this volume. A few tables and figures are found throughout the book.

Karlsson, Svante. **Oil and the World Order: American Foreign Oil Policy.** Totowa, NJ: Barnes & Noble, 1986, 308 p. ISBN 0-389-20663-6.

This volume is divided into three parts, each defining a certain period of development in the international oil market under U.S. controls. Part 1, "American Challenge," focuses on U.S. oil interests and their efforts to take control and shape the international oil market. Part 2, "The Hegemony," discusses the period under which U.S. oil interests had control over the international oil market, with disturbances caused by such incidences as the Iran crisis and the Suez conflict. OPEC was created during this period OPEC. Part 3, "The Wings of Change," analyzes the development of the international oil market during the very turbulent period in the early 1970s. Statistical information is found in tables throughout the book. Appendices provide the world oil production for the periods, 1860–1915 and 1920–1980. The bibliography contains books, articles, and official documents.

Melosi, Martin V. **Coping with Abundance: Energy and Environment in Industrial America.** Philadelphia, PA: Temple University Press, 1985, 355 p. ISBN 0-87722-372-6.

Melosi examines the United States as a producer and consumer of energy from the 1820s Industrial Revolution through the energy crisis in the 1970s. Part one discusses the coal age, part two deals with oil to produce energy, part three discusses the energy-intensive society, and part four examines the period of scarcity. In the epilogue Melosi treats the energy problem of the 1980s. A list of books for further reading is found at the end of each chapter. A few tables are found throughout the book.

Riva, Joseph P., Jr., John J. Schanz, and John G. Ellis. **U.S. Conventional Oil and Gas Production: Prospects to the Year 2000.** Westview Special

Studies in Natural Resources and Energy Management. Boulder, CO: Westview Press, 1985, 145 p. ISBN 0-8133-7066-3.

The authors wrote this book to show the status of two very important resources (oil and gas) to supply U.S. energy needs. Oil and gas not provided by domestic reserves will have to be obtained from oil and gas exporting countries. The future availability of both oil and gas can only be estimated. The first part of this book discusses oil production to the year 2000. The second part deals with natural gas production to the year 2000. The book concludes by giving some observations. The book has many tables, and each part has a few notes.

Tétreault, Mary Ann. **Revolution in the World Petroleum Market.** Westport, CT: Quorum Books, 1985, 271 p. ISBN 0-89930-012-X.

The author, who spent some time in Kuwait and Bahrain, looks at the evolution of the international oil industry in the twentieth century and shows ways governments and companies adjusted to the oil revolution. Adjustments made to the higher prices of crude oil are discussed. The author then discusses how the strategies of major oil companies, oil importing and oil exporting countries have altered their roles in the international oil market. The book has several tables and a list of abbreviations used. There is a short essay on sources used.

Tsai, Hui-Liang. **Energy Shocks and the World Economy: Adjustment Policies and Problems.** New York: Praeger Publishers, 1989, 199 p. ISBN 0-275-93192-7.

This book is written at a level for a well-read individual. It gives an overview of the economic aspects of energy issues and problems. It also gives energy policies as applied in response to the energy shocks of the 1970s. The recycling process of the 1970s is appraised. Energy was always considered essential but abundant and not expensive, so the shocks and reactions caused by the sudden increases in international oil prices caused an upheaval in the world economy in the 1970s. The book has notes and a lengthy bibliography. Most chapters have several tables as explanatory information.

Uslander, Eric M. **Shale Barrel Politics: Energy and Legislative Leadership.** Stanford Studies in the New Political History, Stanford, CA: Stanford University Press, 1989, 241 p. ISBN 0-8047-1703-6.

Uslander served as a consultant to the Select Committee on Committees of the House of Representatives in 1979–1980 and witnessed energy politics firsthand. The purpose of the committee was to establish an Energy Committee in the House of Representatives, but it had no success. His work on the Select Committee led to the framework for this book. The book is about energy politics and focuses on the conditions for cooperation and conflict in a legislative body. Chapters are introduced by the use of quotations from Ben Jonson's plays "Everyman in His Humour" and "Everyman Out of His Humour." A few tables are interspersed throughout the book. There are several pages of bibliographic references.

Vergara, Walter, Nelson E. Hay, and Carl W. Hall, eds. **Natural Gas: Its Role and Potential in Economic Development.** Westview Special Studies in Natural Resources and Energy Management. Boulder, CO: Westview Press, 1990, 319 p. ISBN 0-8133-7812-5.

This book deals with the use of natural gas as an underutilized energy source. Natural gas is an abundant resource that should be used more effectively. It was looked at as the unwanted child of oil exploration. The authors show the potential gas has as an energy producer in industry. The book examines the cost and availability of natural gas. The use of gas in the fertilizer industry and transportation are discussed. References and notes are found with the chapters. Several figures and tables are used to illustrate the book. Terms and conversion factors and abbreviations and acronyms are found in the appendices.

Weimer, David Leo. **The Strategic Petroleum Reserve: Planning Implementation and Analysis.** Contributions in Economics and Economic History, No. 48. Westport, CT: Greenwood Press, 1982, 229 p. ISBN 0-313-23404-3.

The book begins by giving the origin of the Strategic Petroleum Reserve Program (SPR) to help protect us when there is an oil embargo. In 1973 the Arab oil embargo and reduced oil production by Iran caused sharp rises in the world price of oil, thus affecting economic activity. Petroleum stockpiles during market disruptions can reduce economic losses by moderating price rises and reducing transfer of wealth from oil importers to oil exporters. The Ford, Carter, and Reagan administrations pursued the development of crude oil stockpiles owned and controlled by the government. These SPR reserves are stored in salt formations in Louisiana and Texas, but they have not met the goals. More appropriate stockpiling policies are suggested. The book has a list of abbreviations, figures, and tables. An appendix gives the SPR provisions of the Energy Policy and Conservation Act, Public Law 94-163, December 22, 1978. There is also a lengthy bibliography.

Coal Industry

Banks, Ferdinand E. **The Political Economy of Coal.** Lexington, MA: Lexington Books, 1985, 270 p. ISBN 0-669-06169-7.

Banks wrote this book on applied energy economies, stressing coal, discussing the production, reserves, supply, demand, price, trade, gasification, and effects on the environment. He ends the book by giving a summary of the world coal market. There is one chapter on oil, one on electricity, and one on nuclear energy. The book has tables and figures and includes a bibliography on energy literature.

Calzonetti, Frank J., and Mark S. Eckert. **Finding a Place for Energy: Siting Coal Conversion Facilities.** Resource Publications in Geography. Washington, DC: Association of American Geographers, 1981, 70 p. ISBN 0-89291-147-6.

This small volume, dealing with coal, examines the energy siting problem and the many issues that must be considered when choosing the proper location for an energy facility. Many figures are interspersed throughout the book. A bibliography of books, journal articles, and government documents is included.

Gulliford, Andrew. **Boomtown Blues: Colorado Oil Shale, 1885–1985.** World Resources and Environmental Issues Series. Niwot, CO: University Press of Colorado, 1989, 302 p. ISBN 0-87081-178-9.

The West is an area of boomtowns that mostly became ghost towns. The first oil shale boom produced no boomtowns, had subtle changes, and finally went bust. Towns boomed because they were adjacent to oil shale, coal, oil, and natural gas. Oil shale is found in the Rocky Mountains and there was hope it could be used for energy during the energy crisis of the 1970s. Boomtown expansion caused property values to skyrocket. Changes took place rapidly with towns growing overnight. This study focuses on the history of the towns of Rifle, New Castle, Silt, Parachute, and Grand Junction during the boom and after the bust. The book has several pages of endnotes, primary sources, and secondary sources. Many interesting maps and pictures are found throughout.

Harris, Richard A. **Coal Firms Under the New Social Regulation.** Duke Press Policy Studies. Durham, NC: Duke University Press, 1985, 188 p. ISBN 0-8223-0648-4.

The signing of the Surface Mining Control and Reclamation Act (SMCRA) in 1977 was a victory for environmentalists and a defeat for the coal industry according to those who supported the Act. This volume develops several themes, such as the development of coal regulatory policies, the political-economic framework that affects coal utilization, the coal industry and firm characteristics, the Surface Mining Control and Reclamation Act and firm preference, and the social outcomes of government regulation. A few tables are interspersed throughout the text. Bibliographic references are found at the end of the book.

Rogers, Kathryn S. **U.S. Coal Goes Abroad: A Social Action Perspective on Interorganizational Networks.** Praeger Special Studies, Praeger Scientific. New York: Praeger Publishers, 1986, 159 p. ISBN 0-03-004354-9.

Rogers tells the story of how the coal industry in the United States reacted to the increased demand for exports during the period 1980 to 1982. It was written for scientists, sociologists, and government-business policymakers. The author discusses the interorganizational patterns that occurred in the U.S. coal industry due to the increased demand for exporting steam coal and the effects on U.S. coal firms. Chapters have bibliographic notes. There is a theoretical appendix followed by a lengthy bibliography.

Zimmerman, Martin B. **The U.S. Coal Industry: The Economics of Policy Choice.** MIT Press Energy Laboratory Series, 3. Cambridge, MA: MIT Press, 1981, 205 p. ISBN 0-262-24023-0.

The work for this book was supported by grants from the Council on Wage and Price Stability, the Department of Energy, and the Center for Energy Policy Research at the Massachusetts Institute of Technology. The book begins with a discussion of the coal environment. Models are developed to show coal supply, transportation, and demand. A variety of forces affect the development of the coal industry. A chapter is devoted to showing how supply influences coal costs, production, and distribution patterns. The coal policy has very little impact on oil imports. References appear at the end of each chapter. Notes are found for each chapter. There are four appendices and many tables throughout the book.

Nuclear Energy

Babin, Ronald. **The Nuclear Power Game.** Translated by Ted Richmond. Montreal: Black Rose Books, 1985, 236 p. ISBN 0-920057-30-6.

This book gives the history of the internal dynamics of the Canadian atomic power industry. The second part analyzes the nature of the opposition to the industry. The antinuclear movement developed in two stages. The first stage, lasting from the beginning of the 1970s to the 1979 accident at Three Mile Island, was one of rapid growth. In the second stage, dating from the Three Mile Island accident, the movement adopted a more flexible attitude and tried to become a political force by seeking alliances with other progressive social movements.

Ball, Howard. **Justice Downwind: America's Atomic Testing Program in the 1950s.** New York: Oxford University Press, 1986, 289 p. ISBN 0-19-503672-7.

The entire book tells the story of the atom, the making and testing of nuclear weapons, along with the dangers involved. It tells about the impact nuclear power had on our social system. The book examines the various conflicts—medical, scientific, legal, and political—and tells what happened as the United States entered the Atomic Age in the 1940s. Many special studies were made to show the association between radiation exposure and cancer. The book, written in an easy reading style, has three appendices, notes, and a bibliography. There are several illustrations.

Campbell, John L. **Collapse of an Industry: Nuclear Power and the Contradictions of U.S. Policy.** Cornell Studies in Political Economy. Ithaca, NY: Cornell University Press, 1988, 231 p. ISBN 0-8014-2111-X.

This book tells the story of commercial nuclear power and its collapse as an industrial enterprise in the United States. In the book are figures and tables giving additional information for chapters. A few footnotes are found throughout. Several pages give bibliographic references.

Freudenburg, William R., and Eugene A. Rosa, eds. **Public Reactions to Nuclear Power: Are There Critical Masses?** AAAS Selected Symposium 93. Boulder, CO: Westview Press, 1984, 370 p. ISBN 0-86531-708-9.

Based on a symposium held at the January 3–8, 1982, American Association for the Advancement of Science (AAAS) annual meeting in Washington, D.C., sponsored by the Rural Sociological Society, the American Sociological Association, and AAAS. The book deals with the marches and demonstrations opposing the nuclear power industry. It is written in a readable style, giving public attitudes toward nuclear power.

Jasper, James M. **Nuclear Politics: Energy and the State in the United States, Sweden, and France.** Princeton, NJ: Princeton University Press, 1990, 327 p. ISBN 0-691-07841-6.

This book began as a dissertation where the author conducted a case study of the controversy over the Diablo Canyon reactor. He begins by explaining nuclear policies in part one. In the second part he discusses creating nuclear systems: the triumph of technological enthusiasm, 1960–1973. In part three he discusses dilemmas arising from public opposition and the oil crisis, 1973–1976. Part four examines high costs and decentralization of control in the United States. Only chapters dealing with the United States were examined. Finally, the author gives his conclusions and what we have learned. The book has a list of tables, figures, and technical and international abbreviations. The book has a lengthy bibliography and includes many personal interviews, so there is a list of informants.

Lovins, A. B., L. H. Lovins, and Leonard Ross. **Energy/War: Breaking the Nuclear Link.** New York: Harper & Row, 1981, 164 p. ISBN 0-06-090852-1.

This book has eleven chapters dealing with such topics as nuclear power and nuclear bombs, disguises and safeguards, the collapse of nuclear power, how to save oil, the fatuity of U.S. policy, implementing nonnuclear futures, the nuclear arms race, nuclear power and developing countries, and, in the final chapter, a once-in-a-lifetime opportunity. Nuclear power is the slowest, most costly, and most dangerous way to displace oil. A condensation of this book appeared in **Foreign Affairs,** Summer 1980.

Murray, Raymond L. **Nuclear Energy,** 3rd ed. Pergamon Unified Engineering Series, Vol. 22. New York: Pergamon Press, 1988, 350 p. ISBN 0-08-031628-X.

Provides a factual description of basic nuclear phenomena. Not only are the devices and processes described, but attention is also given to the problems and opportunities that are inherent in a nuclear age. The material is designed for those who want to know about the role of nuclear energy in our society. The basic concepts are presented in a "real world" context. The sequences of presentation proceed from fundamental facts and principles through a variety of nuclear devices to the relation between nuclear energy and peaceful applications.

Ott, Karl O., and Bernard I. Spinard, eds. **Nuclear Energy: A Sensible Alternative.** New York: Plenum Press, 1985, 386 p. ISBN 0-306-41441-4.

Many contributors wrote chapters for the compilation of this book on nuclear energy. They attempted to provide in-depth articles on the use of nuclear energy for peaceful domestic purposes. Some of the topics covered are energy and society, the economics of nuclear power, recycling and proliferation, and risk assessment. Tables and figures are used as illustrations. References are found at the end of each chapter.

Platt, A. M., J. V. Robinson, and O. F. Hill, eds. **Nuclear Fact Book,** 2nd ed. London: Harwood Academic Publishers, 1985, 192 p. ISBN 3-7186-0273-3.

The *Nuclear Fact Book* provides concise summaries and essential facts in the areas of energy production, costs, consumption, nuclear energy production, fuel cycle, and wastes. This is a guide to waste management costs and identifies high-level radioactive wastes and regulations on transportation of nuclear materials.

Shrader-Frechette, K. S. **Nuclear Power and Public Policy: The Social and Ethical Problems of Fission Technology.** Boston, MA: D. Reidel, 1980, 176 p. ISBN 90-277-1054-6.

This book grew out of projects funded by the Kentucky Humanities Council and the Environmental Protection Agency. It gives the history of using atomic fission to generate electricity and describes the principal features of a nuclear reactor. The author assumes that reactors will operate normally, but questions how the radioactive waste can be stored without causing problems. A core-melt catastrophe is discussed showing why public policy regarding liability coverage violates constitutional principles. The author evaluates the cost benefits of nuclear technology and claims that, economically, it is both illogical and unethical. The last chapter deals with nuclear safety and the naturalistic fallacy. Policy-making procedures that should be followed in dealing with nuclear technology are outlined.

Smart, Ian, ed. **World Nuclear Energy: Toward a Bargain of Confidence.** Baltimore. MD: Johns Hopkins University Press, 1982, 394 p. ISBN 0-8018-2652-7.

Prepared under the auspices of the Royal Institute of International Affairs and based on material published by the International Consultative Group on Nuclear Energy (ICGNE), which no longer exists, this book is a composite of some work done by the ICGNE group. Such topics as peaceful nuclear relations, nuclear energy and international cooperation, nuclear nonproliferation, international safeguards, viability of the civil nuclear industry, and international custody of plutonium stocks are discussed. Many figures and tables are presented. A lengthy glossary and abbreviations are included.

Union of Concerned Scientists. **The Nuclear Fuel Cycle.** MIT Press Environmental Studies Series. Cambridge, MA: MIT Press, 1975, 291 p. ISBN 0-262-21005-3.

A worldwide controversy has developed over the safety, environmental, and national security impact of the expanding program of utilizing nuclear reactors to generate electricity. This volume presents fundamental information on the development of nuclear power in an essentially nontechnical form. Topics discussed include radiation-induced lung cancer among uranium miners, radiation hazards associated with uranium mill operations, catastrophic nuclear reactor accidents, nuclear safeguard problems, transportation of radioactive wastes, nuclear fuel reprocessing, and storage and disposal of high-level radioactive wastes.

Renewable Energy

Antonopoulos, A. A., ed. **Biotechnological Advances in Processing Municipal Wastes for Fuels and Chemicals.** Park Ridge, NJ: Noyes Data Corporation, 1987, 488 p. ISBN 0-8155-1122-1.

This book is based on a 1984 symposium held in Minneapolis, Minnesota, dealing with the conversion of waste for fuels and chemicals. The authors represented many fields, but the sections on the biological production of fuels and chemicals and the generation and extraction of landfill gas are of special interest to those interested in energy. This information should be useful to researchers, policy makers, and entrepreneurs. The book has a few tables and figures. References are found at the end of each chapter.

Blackburn, John O. **The Renewable Energy Alternative: How the United States and the World Can Prosper Without Nuclear Energy or Coal.** Durham, NC: Duke University Press, 1987, 201 p. ISBN 0-8223-0687-5.

Blackburn discusses how the United States and the world can prosper without nuclear energy or coal. Some of the renewable energy sources he discusses are solar energy, waterpower, wind power, photovoltaic electricity, geothermal electricity, and industrial cogeneration with biomass fuels. The book has a few figures and tables. There are extensive bibliographic notes for each chapter as well as a bibliography.

Flavin, Christopher. **Energy and Architecture: The Solar Conservation Potential.** Worldwatch Paper, 10. Washington, DC: Worldwatch Institute, 1980, 63 p. ISBN 0-916468-39-3.

The energy problem has many facets—transportation, manufacturing, and agriculture depend on fossil fuels. Also dependent on fossil fuels is the heating, cooling, and lighting of buildings so, in a way, it is an architectural problem. This small volume deals with the architectural aspect of energy, with plans to build to save energy. There are bibliographic notes of 85 entries.

————. **Wind Power: A Turning Point.** Worldwatch Paper, 45. Washington, DC: Worldwatch Institute, 1981, 56 p. ISBN 0-916468-44-5.

Wind power is a practical and substantial source of electricity and mechanical power. With an expanded use of hydropower and the development of new renewable sources of energy, such as photovoltaic cells, much of our electricity needs can be met. This author examines ways in which the wind can be harnessed to produce energy. There are 81 bibliographic notes.

Flavin, Christopher, Rick Piltz, and Chris Nichols. **Sustainable Energy.** Washington, DC: Renew America, 1989, 48 p. No ISBN.

There is a wealth of information about renewable energy sources in this small volume. The authors begin by stating that we must redirect U.S. energy policy to prevent catastrophic climate change. Fuel economy, energy-wise buildings, lighting and appliances, wind power, industrial efficiency, solar thermal power, photovoltaic electricity, geothermal electricity, ocean power, hydropower and biomass energy are some of the topics discussed. They also examine the use of ethanol and hydrogen as transportation fuels. They summarize by stating that the nation's energy policy must also become a climate protection strategy and we must rely on renewable energy sources.

Gray, Charles L., Jr., and Jeffrey A. Alson. **Moving America to Methanol: A Plan to Replace Oil Imports, Reduce Acid Rain, and Revitalize Our Domestic Economy.** Ann Arbor, MI: University of Michigan Press, 1985, 144 p. ISBN 0-472-10071-8.

Petroleum supply is one of the United States' energy problems. The book begins by discussing acid rain and energy problems, where the strategy is to limit the amount of SO_2 that can be emitted from a given power plant using high-sulfur coal and require the use of scrubbers by plants using low-sulfur coal. This would be very costly, so the author suggests the energy problem be solved by using an alternative transportation fuel, such as methanol, which can be produced from domestic resources. After discussing the overall effects of a methanol program the author provides a blueprint for a transition to the use of methanol. The book has tables and figures, and bibliographic notes for each chapter.

Hasselriis, Floyd. **Refuse-Derived Fuel Processing.** An Ann Arbor Science Book. Stoneham, MA: Butterworth Publishers, 1984, 340 p. ISBN 0-250-40314-5.

The author wrote this book because our society is suffering from pollution from our wastes, limited space, and limited resources. He endeavors to investigate and solve these problems. He begins by explaining just what municipal solid waste is and reviews the process of handling refuse from the curbside to the processing plant and the complex processing system. He then goes into the process of reducing the size of refuse by shredding it as it goes into the separation process. In this process refuse is separated as to size, glass, aluminum, tin, etc. Seven appendices are found in the book. Each chapter has references. Figures are found throughout the book.

National Research Council. Committee on Production Technologies for Liquid Transportation Fuels. **Fuels to Drive Our Future.** Washington, DC: National Academy Press, 1990, 223 p. ISBN 0-309-04142-2.

This report was prepared by a committee at the request of the U.S. Department of Energy when there was a decline in the fraction of U.S. transportation fuels supplied by domestic resources. The purpose of the study is to outline programs for producing liquid transportation fuels from domestic resources, with cost reduction and environmental problems caused by the use of transportation fuels as major issues to keep in mind. Major conclusions and recommendations of the committee are given. There are eleven appendices, a glossary, references, and bibliography.

Parker, Colin, and Tim Roberts, eds. **Energy from Waste: An Evaluation of Conversion Technologies.** New York: Elsevier Applied Science Publishers, 1985, 217 p. ISBN 0-85334-352-7.

This book is an evaluation of technologies used to recover energy from municipal solid waste. Such technologies as the biological processes, refuse-derived fuel manufacture and combustion, and pyrolysis/gasification are discussed at length. The book has tables and figures used as illustrations. There are references and four appendices.

Pleatsikas, Christopher J., Edward A. Hudson, and Richard Goettle, IV. **Solar Energy and the U.S. Economy.** A Westview Replica Edition. Boulder, CO: Westview Press, 1982, 152 p. ISBN 0-86531-326-1.

The authors discuss the economic effects of investing in solar energy technologies, examining the cost of the technologies, the cost of fossil fuels, and the amount of solar energy that can be supplied. Using an economic model they project the economic growth, inflation, productivity, and capital investment for the period 1980–2000. They point out potential economic effects of large solar energy investments that must be considered in the future energy plans of the United States. An appendix gives background material and references for Chapter 2. Each chapter has several explanatory tables.

Porteous, Andrew. **Refuse Derived Fuels.** Energy from Wastes Series. New York: Halsted Press, 1981, 137 p. ISBN 0-470-27170-1.

The author reminds us of the escalation in energy costs and its effects. He explores refuse derived fuels, their production, use, environmental aspects, and economic factors. Porteous calls anything for which there is no further use "waste." Bibliographic references are found at the end of each chapter. The book is illustrated with both figures and tables.

————. **Renewable Energy in Cities.** New York: Van Nostrand Reinhold, 1984, 376 p. ISBN 0-442-21654-8.

Work for this book was sponsored by the Department of Energy. It entailed a study of renewable energy, including proper energy planning by city energy officials, a study of land-use patterns, energy supply and distribution, and the link between energy, housing, employment, and economic development. A problem was to decide if a particular energy technology was suitable for a particular neighborhood. Liquid renewables used for transportation in cities are ethanol and methanol. Bibliographic notes are given for each chapter and a bibliography for additional reading. Tables are found in each chapter.

Rogoff, Marc Jay. **How to Implement Waste-to-Energy Projects.** Park Ridge, NJ: Noyes Publications, 1987, 202 p. ISBN 0-8155-1132-9.

Rogoff wrote this book to be used by those responsible for solid waste disposal since many communities have no answer to the disposal of the ever-mounting increase in their solid waste disposal needs. Due to restrictions on siting and operation of sanitary landfills, waste-to-energy is a good alternative. Waste-to-energy projects aim to dispose of solid waste in an economically efficient and environmentally acceptable manner. The book has references with each chapter. There are a few tables and a list of acronyms and abbreviations.

Sawyer, Stephen W. **Renewable Energy: Progress, Prospects.** Resource Publications in Geography. Washington, DC: Association of American Geographers, 1986, 102 p. ISBN 0-89291-192-1.

This book was written to show how the world really depends on energy for economic growth and development. Renewable energy resources are linked to geographic factors in many areas. Fossil fuels can provide an ice pavilion in Abidjan or a tropical greenhouse in St. Paul. Renewable energy often exploits resources inherent to a specific place. Renewable energy is dependent on high technology, such as turning the energy of flowing water or sun into electricity. The author assesses the prospects of renewable energy through the year 2000. He concludes the book with several observations: major changes in technology, public support, the part played by tax policy, utilities influence the rate of renewable energy use, and regional distribution of renewable energy. This small publication has many figures and tables, a glossary, and a bibliography for additional reading.

Shea, Cynthia Pollock. **Renewable Energy: Today's Contribution, Tomorrow's Promise.** Worldwatch Paper, 81. Washington, DC: Worldwatch Institute, 1988, 68 p. ISBN 0-916468-82-8.

Shea discusses the use of renewable energy sources used as alternatives to dwindling oil supplies. Some of the renewable energy sources she examines are wind, solar energy, biomass, and hydropower. She uses tables to show the energy that can be obtained from those sources. There is a bibliography of 116 items for additional reading.

Sorensen, Bent. **Renewable Energy.** Energy Science and Engineering: Resources, Technology, Management. New York: Academic Press, 1979, 683 p. ISBN 0-12-656150-8.

The study of renewable energy sources is an emerging science. Earlier, more effort was devoted to the extraction and utilization of nonrenewable energy resources than to renewable ones. The purpose of this book is to provide information on problems dealing with renewable energy. The author begins by discussing the nature and origin of renewable energy and tells what the renewable sources are, such as solar radiation, wind, ocean waves, tides, geothermal flows, and nuclear energy. He then describes energy conversion devices, means of transportation and storage, and the virtues of renewable energy resources. The book has many illustrative figures and several pages of references.

Sperling, Daniel. **New Transportation Fuels: A Strategic Approach to Technological Change.** Berkeley, CA: University of California Press, 1988, 532 p. ISBN 0-520-06087-3.

This book begins with a discussion of the transportation energy problem by reviewing the past and preparing for the future. With the possibility of petroleum transportation fuels being phased out he discusses some of the possible replacements for petroleum and the part the government should play in the transition to those fuels used as a substitute. Biomass, oil shale and oil sands, ethanol, and hydrogen fuel are some possible substitutes that are discussed. The book has many figures and illustrations. Notes with bibliographic references are found at the end of the book.

Sperling, Daniel, ed. **Alternative Transportation Fuels: An Environmental and Energy Solution.** Westport, CT: Quorum Books, 1989, 326 p. ISBN 0-89930-407-9.

Many authors contributed to this volume on alternative fuels. Legislators and policy-makers make many mistakes simply because they do not understand the energy market and the implications of selecting a particular option. One of the great concerns is the use of petroleum for powering motor vehicles. The book is the result of a symposium held July 17–19, 1988, in Monterey, California, dealing with alternative transportation fuels in the 1990s and beyond. This volume should give a broader understanding of transportation fuel choices and sound energy and transportation policies regarding clean-burning alternative transportation fuels. Most chapters have conclusions and recommendations along with bibliographic notes. The book has many figures and tables.

Swan, Christopher C. **Suncell: Energy, Economy and Photovoltaics.** San Francisco, CA: Sierra Club Books, 1986, 231 p. ISBN 0-87156-751-2.

This most interesting book is devoted to the miraculous invention of photovoltaics. Originally developed for consumer products by RCA in the late 1950s, photovoltaics really didn't become popular until the 1960s when NASA needed lightweight, long-term sources of electricity to operate satellites. Solar Technology International, the real beginning of the photovoltaic industry, was founded in 1975. At that time Exxon bought the Solar Power Corporation. Since then much research and development work was done on photovoltaics. It produces electrical generation with no noise, no air pollution, no international dependence, and no radioactivity. A minor problem does exist: the toxic chemicals used in producing silicon for photovoltaics. Coal, oil, nuclear, and hydroelectric as sources of energy have been subsidized by the government, but it does not look favorably toward renewables such as photovoltaics. The book has a small bibliography of books, government publications, and magazines, as well as many illustrations.

Articles and Government Documents

Energy Policy

General

Berg, Sanford V., and Prakash Lougani. **"Frameworks for State Energy Policy: The Case of Florida."** *Strategic Planning Energy and Environment* 11 (Fall 1991): 56–69.

Berry, John M. **"Large Problems Still Loom Over National Energy Policy."** *Financier* 13 (September 1989): 6–10.

Cameron, Lori. **"States Seize Energy Baton Dropped at Federal Level."** *Forum for Applied Research and Public Policy* 5 (Summer 1990): 72–76.

Canes, Michael E. **"Rational United States Energy Policies."** *Middle East Review* 16 (Summer 1984): 45–51.

Capros, Pantelis, and Emmanuel Samouilidis. **"Energy Policy Analysis."** *Energy Policy* 16 (February 1988): 36–48.

Cavanaugh, Ralph, and others. **"Toward a National Energy Policy."** *World Policy Journal* 6 (Spring 1989): 239–264.

Dreyfus, D. A. **"Persistent National Issue."** *Journal of Professional Issues in Engineering* 114 (January 1988): 1–8.

Ender, Richard L., and Choon K. Kim, eds. **"Current Issues in Energy Policy: A Symposium."** *Policy Studies Journal* 13 (December 1984): 303–431.

"Forum: Energy Policy: A Post-War Framework." *CRS (Cong Research Service) Review* 12 (March/April 1991): 1–35.

Holing, Dwight. **"America's Energy Plan: Missing in Action."** *Amicus Journal* 13 (Winter 1991): 12–20.

Idelson, Holly. **"Energy Politics Emerge as Oil Crisis Subsides: Persian Gulf War Provided Impetus for Sweeping Bill, but Competing Interests Put a Drag on the Action."** *Congressional Quarterly Weekly Report* 49 (June 15, 1991): 1569–1574.

James, Patrick, and Robert Michelin. **"The Canadian National Energy Program and Its Aftermath: Perspectives on an Era of Confrontation."** *American Review of Canadian Studies* 19 (Spring 1989): 59–81.

Kash, Don E., and Robert W. Rycroft. **"Energy Policy: How Failure Was Snatched from the Jaws of Success."** *Policy Studies Review* 4 (February 1985): 433–444.

Kriz, Margaret E. **"Balancing Act: As the Bush Administration Tries Writing a National Energy Strategy, Officials Are Weighing Environmental Concerns Against Renewed Worries About Reliance on Overseas Oil."** *National Journal* 22 (January 20, 1990): 125–128.

Kuntz, Phil. **"Energy: Iraq Hits a Raw Nerve in U.S., but Policy Changes Unlikely; Industry Cites Need for Increased Domestic Drilling; Environmentalists Argue for Alternative Fuels."** *Congressional Quarterly Weekly Report* 48 (August 11, 1990): 2588–2590.

Landsberg, Hans H. **"U.S. Energy Policy in Historical Perspective."** *Resources Policy* 15 (December 1989): 297–308.

New York (State) Energy Office. Bureau of Policy Analysis. *New York State Annual Energy Review: Energy Consumption, Supply, and Price Statistics, 1970–1988.* Albany, NY: 1990, 97 p.

Olsen, Marvin E., and others. **"Public Opinion Versus Governmental Policy on National Energy Issues"** in Braungart, Richard G., ed. *Research in Political Sociology, 1985.* Greenwich, CT: JAI Press, 1985, 189–210.

Patterson, W. **"Energy Issues Another Challenge."** *New Scientist* 121 (January 28, 1989): 45–50.

"Petroleum Markets: Outlook and Policy Implications." *Contemporary Policy Issues* 5 (July 1987): 1–33.

Pirog, Robert, and Stephen C. Stamos. **"Energy Concentration: Implications for Energy Policy and Planning."** *Journal of Economic Issues* 19 (June 1985): 441–449.

Sabatier, Paul A., and Hank C. Jenkins-Smith, eds. **"Special Issue: Policy Change and Policy-Oriented Learning: Exploring an Advocacy Coalition Framework."** *Policy Sciences* (Amsterdam) 21:2/3 (1988): 123–277.

Sampson, Martin W., III. **"Rapid Deployment in Lieu of Energy Policy?"** in Goldman, Joseph R., ed. *American Society in a Changing World.* Lanham, MD: University Press of America, 1987, 288–318.

Smith, Tim R. **"U.S. Energy Policy in a Changing Market Environment."** *Federal Reserve Bank of Kansas City Economic Review* 71 (September/October 1986): 16–30.

Starr, Chauncey. **"National Energy Policy: A Retrospective."** *Energy Systems and Policy* 13:1 (1989): 51–61.

Stern, Paul C. **"Blind Spots in Policy Analysis: What Economics Doesn't Say About Energy Use."** *Journal of Policy Analysis and Management* 5 (Winter 1986): 200–233.

U.S. Congress. House. Committee on Interior and Insular Affairs. Subcommittee on Mining and Natural Resources. *Potential Contribution of Mining and Energy Extractive Industries to the Administration's National Energy Policy: Oversight Hearing, February 22, 1990.* 101st Congress, 2nd session. Washington, DC: GPO, 1991, 491 p.

U.S. Congress. Senate. Committee on Energy and Natural Resources. *Energy Initiatives of the 97th Congress.* 98th Congress, 2nd session. Washington, DC: GPO, 1984, 256 p.

————. *National Energy Policy Act of 1988 and Global Warming: Hearings, August 11–September 20, 1988.* 100th Congress, 2nd session. Washington, DC: GPO, 1989, 543 p.

————. *DOE's National Energy Plan and Global Warming: Hearing, July 26, 1989.* 101st Congress, 1st session. Washington, DC: GPO, 1989, 155 p.

————. *Methanol as a Potential Alternative Fuel in Our Future Energy Policy: Hearing, October 17, 1989.* 101st Congress, 1st session. Washington, DC: GPO, 1990, 361 p.

————. *National Energy Policy Act of 1989 and Federal Energy Management Amendments of 1990: Hearing, April 5, 1990, on S. 324, to Establish a National Energy Policy to Reduce Global Warming and for Other Purposes, and S. 2191, to Amend Part 3, Title V, of the National Energy Conservation Policy Act of 1990.* 101st Congress, 2nd session. Washington, DC: GPO, 1990, 172 p.

————. *Key Elements of a National Energy Policy: Hearing, October 2, 1990.* 101st Congress, 2nd session, Washington, DC: GPO, 1991, 189 p.

U.S. Congress. Senate. Committee on Energy and Natural Resources. Subcommittee on Energy Regulation and Conservation. *Implications of Proposed National Energy Policy Legislation for Natural Gas: Hearing, April 3, 1991, on S. 341 (Title X), to Reduce the Nation's Dependence on Imported Oil, to Provide for the Energy Security of the Nation and for Other Purposes.* 102nd Congress, 1st session. Washington, DC: GPO, 1991, 270 p.

U.S. Department of Energy. *United States Energy Policy, 1980–1988.* Washington, DC: 1988, 175 p.

Uslaner, Eric M. **"Energy Politics in the U.S.A. and Canada."** *Energy Policy* 15 (October 1987): 432–440.

Zelby, Leon W., and B. Groten. **"Guidelines for the Development of a Consistent Long-Range Energy Policy."** *IEEE [Institute of Electric and Electronics Engineers] Technology and Society Magazine* 9 (April 1990): 17–22.

National Security

Adkin, P. **"Mass Burning: Garbage In: Energy Out."** *Mechanical Engineering* 110 (December 1988): 46–51.

Ajami, Riad. **"U.S. Oil Security: An Oil Imports Policy for the Turbulent 1990s."** *Middle East Insight* 7:2/3 (1990): 54–63.

"Are We Slipping into Another Energy Crisis?" *Conservation Foundation Letter* No. 4 (1987): 1–8.

Churney, K. L., and others. **Assessing the Credibility of the Calorific Value of Municipal Solid Waste.** Washington, DC, U.S. Department of Energy, Office of Renewable Technology, Energy from Municipal Waste Division, 1984, 40 p.

Davis, Joseph A. **"Energy Security: How Real Is the Threat?"** *Congressional Quarterly Weekly Report* 45 (May 30, 1987): 1115–1120.

"Energy from Waste: On-Site Heat-Recovery Incineration." *Power* 131 (March 1987): W1–15.

"Energy from Wastes." *Power* 132 (March 1988): W1–37.

Gettinger, Stephen. **"The Energy Crisis May Be Over, but Some in Congress Worry the Nation Is Still Unprepared."** *Congressional Quarterly Weekly Report* 43 (January 12, 1985): 75–80.

Goldstein, G. **"Rechanneling the Waste Stream."** *Mechanical Engineering* 111 (August 1989): 44–50; Discussion. 111 (October 1989): 6.

Hervey, Jack. L. **"Energy Dependence and Efficiency."** *Economic Perspectives* 15 (September/October 1991): 2–21.

Johnson, Debra. **"Where Do All the Tyres Go?"** *Energy Economist* (May 1990): 7–9.

Jordan, R. J. **"The Feasibility of Wet Scrubbing for Treating Waste-to-Energy Flue Gas."** *Journal of the Air Pollution Control Association* 37 (April 1987): 422–430.

Klass, D. L., and C. T. Sen. **"Energy from Waste."** *Chemical Engineering Progress* 83 (July 1987): 46–52.

Knight, A. **"Energy from Waste: Recovering a Throwaway Resource."** *EPRI Journal* 13 (October/November 1988): 26–35.

Kodres, C. A. **"Theoretical Thermal Evaluation of Energy Recovery from Incinerators."** *Journal of Energy Resource and Technology* 109 (June 1987): 79–89.

Korzun, E. A. **"Economic Value of Municipal Solid Waste."** *Journal of Energy Engineering* 116 (April 1990): 39–50.

Marier, D. **"Waste-to-Energy Works: Spread the Word."** *Alternative Sources of Energy* No. 91 (May/June 1987): 49–52.

Penner, S. S. **"Waste Incineration and Energy Recovery."** *Energy* 13 (December 1988): 845–851.

"Refuse to Energy Survey: Current Operation, Future Outlook." *Public Works* 120 (May 1989): 77–82+; 120 (June 1989): 68–72.

Sohn, Ira. **"U.S. Energy Security: Problems and Prospects."** *Energy Policy* 18 (March 1990): 149–161.

Stanfield, Rochelle L. **"Fillerup: With What? As the United States Again Starts Worrying About Rising Oil Imports from the Middle East, Then**

Is Renewed Interest in Switching Energy Sources for Transportation." *National Journal* 19 (June 25, 1987): 1892–1897.

Stevens, P. L., and others. **"Indianapolis Resource Recovery Facility: Community Efforts and Technology Required for a Successful Project."** *Journal of Engineering for Gas Turbines and Power* 112 (January 1990): 31–36.

Stoga, Alan J. **"The United States in a Global Economy: Risks and Challenges."** *Fletcher Forum* 13 (Winter 1989): 1–8.

Tay, J. H. **"Energy Generation and Resources Recovery from Refuse Incineration."** *Journal of Energy Engineering* 114 (December 1988): 107–117.

U.S. Congress. House. Committee on Banking, Finance, and Urban Affairs. Subcommittee on Economic Stabilization. *Conventional Fuels and Energy Security: Hearing, November 8, 1989.* 101st Congress, 1st session. Washington, DC: GPO, 1989, 599 p.

————. *Long-Term Energy Security: Hearings, May 9–June 18, 1989.* 101st Congress, 1st session. Washington, DC: GPO, 1989, 642 p.

————. *Energy Security: Energy Efficiency and National Security: Hearing, October 4, 1989.* 101st Congress, 1st session. Washington, DC: GPO, 1990, 266 p.

————. *Long-Term Energy Security Interests of the United States: Hearing, September 26, 1990.* 101st Congress, 2nd session. Washington, DC: GPO, 1990, 259 p.

————. *U.S. Economic and Energy Security Interests in the Persian Gulf: Hearing, August 7, 1990.* 101st Congress, 2nd session. Washington, DC: GPO, 1990, 197 p.

————. *Long-Term Energy Security Interests of the United States: Hearing, December 11, 1990.* 101st Congress, 2nd session. Washington, DC: GPO, 1991, 745 p.

U.S. Congress. House. Committee on Energy and Commerce. Subcommittee on Energy and Power. *U.S. Energy Security: Hearings, March 18, and 23, 1987.* 101st Congress, 1st session. Washington, DC: GPO, 1987, 237 p.

————. *Strategic Petroleum Reserve and EPCA: Hearing, April 19, 1989.* 101st Congress, 1st session. Washington, DC: GPO, 1990, 182 p.

————. *Strategic Petroleum Reserve: Hearing, May 16, 1990, on H.R. 3193, a Bill to Extend Title I of the Energy Policy and Conservation Act, and for Other Purposes.* 101st Congress, 2nd session. Washington, DC: GPO, 1990, 127 p.

U.S. Congress. House. Committee on Energy and Commerce. Subcommittee on Fossil and Synthetic Fuels. *Energy Emergency Preparedness: Hearings, February 1–21, 1984.* 98th Congress, 2nd session. Washington, DC: GPO, 1984, 477 p.

————. *Energy Security Policy: Hearings, August 29–30, 1984.* 98th Congress, 2nd session. Washington, DC: GPO, 1985, 451 p.

U.S. Congress. Office of Technology Assessment. *U.S. Oil Import Vulnerability: The Technical Replacement Capability.* Washington, DC: GPO, 1991, 136 p.

U.S. Congress. Senate. Committee on Energy and Natural Resources. *Energy Security: Hearing, May 20, 1987.* 100th Congress, 1st session. Washington, DC: GPO, 1987, 195 p.

————. *World Oil Outlook: Hearing, March 26, 1990, to Assess the Outlook for the World Oil Market in the 1990s and Its Implications for U.S. Energy, Economics, and Security Interests.* 101st Congress, 2nd session. Washington, DC: GPO, 1990, 157 p.

————. *National Energy Security Act of 1991: Pt. 1: Hearing, February 21, 1991, on S. 341, to Reduce the Nation's Dependence on Imported Oil, to Provide for the Energy Security of the Nation and for Other Purposes, and on the Administration's National Energy Strategy.* 102nd Congress, 1st session. Washington, DC: GPO, 1991, 573 p.

U.S. Congress. Senate. Committee on Governmental Affairs. *Vulnerability of Telecommunications and Energy Resources to Terrorism: Hearings, February 7–8, 1989.* 101st Congress, 1st session. Washington, DC: GPO, 1989, 396 p.

U.S. Department of Energy. *Energy Security: A Report to the President of the United States.* Washington, DC: GPO, 1987, 240 p.

U.S. Energy Information Administration. Office of Energy Markets and End Use. *Annual Energy Outlook, 1987: With Projections to 2000.* Washington, DC: GPO, 1988, 66 p.

Vir, J. **"Site Design for Resource Recovery Facilities."** *Journal of Energy Engineering* 114 (December 1988): 93–98.

White, Allen L., and Marie Zack. **"Avoided-Cost Pricing of Electricity from Waste-to-Energy Plants: The Regulatory Economics of a Negative Fuel Cost."** *Energy Policy* 17 (August 1989): 370–381.

Yergin, Daniel. **"Energy Security in the 1990s."** *Foreign Affairs* 67 (Fall 1988): 110–132.

Yucel, Mine K., and Carol Dahl. **"Reducing U.S. Oil-Import Dependence: A Tariff, Subsidy, or Gasoline Tax?"** *Federal Reserve Bank of Dallas Economic Review* (May 1990): 17–25.

Consumption

Bleviss, Deborah Lynn. **"Improving Vehicle Fuel Economy: A Critical Need."** *Forum for Applied Research and Public Policy* 5 (Spring 1990): 5–12.

Bleviss, Deborah L., and Peter Walzer. **"Energy for Motor Vehicles."** *Scientific American* 263 (September 1990): 102–109.

Branch, Kristi M., and others. **"The Process of Change in the Social Organization of Communities: Energy Development in Rural Areas,"** in Fear, Frank A., and Harry K. Schwarzweller, eds. *Research in Rural Sociology and Development, 1985.* Greenwich, CT: JAI Press, 1985, 91–107.

Campbell, William A. **"Resource Recovery in North Carolina."** *Popular Government* 52 (Summer 1986): 1–10.

Converse, Richard S., and Gary E. Machlis. **"Energy and Outdoor Recreation: A Review and Assessment of the Literature."** *Leisure Sciences* 8:4 (1986): 391–416.

David, Ged R. **"Energy for Planet Earth."** *Scientific American* 263 (September 1990): 54–62.

Gilland, Bernard. **"Population, Economic Growth, and Energy Demand, 1985–2020."** *Population and Development Review* 14 (June 1988): 233–244.

Holdren, John P. **"Energy in Transition."** *Scientific American* 263 (September 1990): 156–163.

———. **"Population and the Energy Problem."** *Population and Environment* 12 (Spring 1991): 231–255.

Jentleson, Bruce W. **"From Consensus to Conflict: The Domestic Political Economy of East-West Energy Trade Policy [Changing Domestic**

Constraints on U.S. Foreign Policy, Late 1940s–1984]. *International Organization* 38 (Autumn 1984): 625–660.

Marcus, Alfred A., and Allen M. Kaufman. **"Why Is It Difficult to Implement Industrial Policies: Lessons from the Synfuels Experience."** *California Management Review* 28 (Summer 1986): 98–114.

Martin, J. M. **"Energy and Technological Change: Lessons from the Last Fifteen Years."** *STI Review* (July 1990): 9–34.

New Jersey. Department of Energy. *The New Jersey Energy Master Plan, December 11, 1985.* Newark, NJ: [n.d.], 253 p.

"Power Generation: Rapid Industrialisation and Population Growth Rates Have Placed Unprecedented Demands on Power Supplies in Developing Countries." *South* (May 1988): 85+. 11-page section.

Renner, Michael G. **"Shaping America's Energy Future."** *World Policy Journal* 4 (Summer 1987): 383–414.

Throgmorton, James A. **"Community Energy Planning: Winds of Change from the San Gorgonio Pass."** *American Planning Association Journal* 53 (Summer 1987): 358–367.

U.S. Congress. House. Committee on Science and Technology. Subcommittee on Transportation, Aviation and Materials. *Methane-Fueled Vehicle RD&D: Hearing, April 26, 1984.* 98th Congress, 2nd session. Washington, DC: GPO, 1984, 88 p.

U.S. Congress. Office of Technology Assessment. *Energy Use and the U.S. Economy.* Washington, DC: GPO, 1990, 65 p.

U.S. Energy Information Administration. Office of Energy Markets and End Use. *Profiles of Foreign Direct Investment in U.S. Energy, 1987.* Washington, DC: 1988, 36 p.

Walker, William. **"Information Technology and the Use of Energy."** *Energy Policy* 13 (October 1985): 458–476.

Economic

General

Benson, Bruce L. **"Boom Towns in the West: Is Profit the Culprit? Most Boom Town Residents Indicate a High Degree of Satisfaction with**

Their Situations." *Review of Regional Economics and Business* 11 (April 1986): 27–38.

Canto, Victor A., and Charles W. Kadlec. **"The Shape of Energy Markets To Come."** *Public Utilities Fortnightly* 117 (January 9, 1986): 21–28.

Coleman, James W. **"Law and Power: The Sherman Antitrust Act and Its Enforcement in the Petroleum Industry."** *Social Problems* (Soc Study Social Problems) 32 (February 1985): 264–274.

Hall, Jane V. **"Financial Fragility and Restructuring of the U.S. Petroleum Industry."** *Economic Forum* 16 (Winter 1986/1987): 69–80.

Miller, John R. **"Petroleum Economics and Management."** *Journal of Energy and Development* 10 (Spring 1985): 165–172.

Ross, Alexander G. **"Corporate Mergers: What Do You Do with the Benefit Plans?"** *Compensation and Benefits Management* 5 (Winter 1989): 145–150.

Schmidt, James R. **"The Distribution of Royalty Interests in Crude Oil Production."** *Southwestern Review of Management and Economics* 4 (Spring 1985): 7–14.

"Social and Private Costs of Alternative Energy Technologies." *Contemporary Policy Issues* 8 (July 1990): 1–307.

U.S. Congress. Senate. Committee on the Judiciary. *Oil Merger Activity: Hearing, March 15, 1984, on Reasons for Current Merger Activity Within the Oil Industry and Assess Their Effect on the Nation.* 98th Congress, 2nd session. Washington, DC: GPO, 1984, 249 p.

U.S. Energy Information Administration. Office of Energy Markets and End Use. *Profiles of Foreign Direct Investment in U.S. Energy, 1988.* Washington, DC: 1989, 36 p.

"Where Have All the Billions Gone?" *Alaska Review of Social and Economic Conditions* 24 (February 1987): 1–35.

Price

Allen, Zachariah, and Thomas D. Mullins. **"Fuel Prices: What Comes Down Can Go Up."** *Public Utilities Fortnightly* 188 (November 13, 1986): 15–22.

Bernard, Martin J., III, and Danilo J. Santini. **"Enhancing Energy Price Stability Through Consistent and Balanced Energy Technology."** *Energy Policy* 17 (December 1989): 554–566.

Berrie, T. W., ed. **"Special Issue: Spot Pricing."** *Energy Policy* 16 (August 1988): 334–408.

Bina, Cyrus. **"Competition, Control and Price Formation in the International Energy Industry."** *Energy Economics* 11 (July 1989): 162–168.

Brown, Stephen P. A., and Keith R. Phillips. **"The Effects of Oil Prices and Exchange Rates on World Oil Consumption."** *Federal Reserve Bank of Dallas Economic Review* (July 1984): 13–21.

Considine, T. J., and T. D. Mount. *Energy Pricing, Employment and Economic Growth: An Econometric Analysis of the New York State Economy.* Ithaca, NY: New York (State) Agriculture Experiment Station, 1984, 29 p.

Desprairies, P. C., and others. **"Progressive Mobilization of Oil Resources: A Factor in Ensuring Moderate Price Rises."** *Energy Policy* 13 (December 1985): 511–523; 14 (February 1986): 15–23.

Fried, Neal, and others. **"The Oil Patch Slide: How Alaska's Economy Compares to Other Oil States."** *Alaska Economic Trends* 6 (December 1986): 1–10.

Gibbons, Elizabeth, and Gerald F. Halpin. **"Import Price Declines in 1986 Reflected Reduced Oil Prices; Despite Price Increases for Major Product Categories Resulting from the Decline in the Dollar, the Overall Import Index Decreased for the Fourth Consecutive Year."** *Monthly Labor Review* 110 (April 1987): 3–17.

Goldstein, Walter. **"Price and Prospects for Oil: Carrying the Burden of Excess Capacity in the 1980s and Beyond."** *Energy Policy* 13 (December 1985): 524–534.

Hafer, R. W. **"Examining the Recent Behavior of Inflation [Impact of Recent Changes in Food and Energy Prices on Inflation Rates]."** *Federal Reserve Bank of St. Louis Review* 66 (August/September 1984): 29–39.

Huntington, Hillard G. **"The U.S. Dollar and the World Oil Market."** *Energy Policy* 14 (August 1986): 299–306.

Junk, Virginia W., and others. **"Impact of Energy Audits on Home Energy Consumption."** *Journal of Consumer Studies and Home Economics* 11 (March 1987): 21–38.

Knapp, Debra R., and others. **"Home Energy Assistance: Actual Cost Should Be the Basis of Payment."** *Public Utilities Fortnightly* 118 (September 18, 1986): 28–33.

Leslie, George A. **"The United States Oil and Gas Industry 1980–1983: The Effects of Price Misjudgement."** *Three Banks Review* (September 1984): 37–53.

Lyons, Paul. **"What They Really Mean Is—Get Out There and Sell."** *Energy Economist* (March 1989): 6–9.

Olsen, Randall J. **"Price Controls, Price Discrimination and the Market for Petroleum [U.S., 1971–1980]."** *Policy Studies Journal* 13 (September 1984): 55–66.

"Petroleum Markets: Outlook and Policy Implications." *Contemporary Policy Issues* 5 (July 1987): 1–33.

Roberts, John. **"The Effect of the Oil Price Collapse on the Gulf Cooperation Council Economies."** *Journal of Energy and Development* 12 (Autumn 1986): 103–114.

Schmidt, Ronald H., and Jeffrey W. Gunther. **"Distributional Implications Reducing Interstate Energy Price Differences."** *Federal Reserve Bank of Dallas Economic Review* (November 1986): 1–13.

Schroeder, Richard C. **"Oil Prices: Collapse and Consequences."** *Editorial Research Reports.* (April 4, 1986): 247–264.

Smith, Tim R. **"Financial Stress in the Oil Patch: Recent Experience at Energy Banks."** *Federal Reserve Bank of Kansas City Economic Review* 79 (June 1987): 9–23.

Trehan, Bharat. **"Oil Prices, Exchange Rates and the U.S. Economy: An Empirical Investigation."** *Federal Reserve Bank of San Francisco Economic Review* (Fall 1986): 25–43.

U.S. Congress. House. *Naval Petroleum Reserves: Hearing, June 11, 1986, Before the Subcommittee on Fossil and Synthetic Fuels of the Committee on Energy and Commerce and the Investigations Subcommittee of the Committee on Armed Services; on H.R. 4843 and H.R. 4859, Bills to Amend the Energy Policy and*

Conservation Act to Improve Policy, Production, and Pricing Mechanisms of the Naval Petroleum Reserves. 99th Congress, 2nd session. Washington, DC: GPO, 1986, 158 p.

U.S. Congress. Office of Technology Assessment. *U.S. Oil Production: The Effect of Low Oil Prices.* Washington, DC: GPO, 1987, 24 p.

U.S. Department of the Interior. Office of Policy Analysis. *Managing Oil and Gas Resources in an Era of Price Instability.* Washington, DC: 1988, 5 p.

U.S. Energy Information Administration. Office of Energy Markets and End Use. *State Energy Price and Expenditure Report, 1986.* Washington, DC: GPO, 1988, 239 p.

Taxation

Testa, William A. **"State Taxation of Energy Production: Regional and National Issues [How Energy Tax Revenues, Particularly Severance Tax Revenue, Redistribute Regional Income in the United States]."** *Economic Perspectives* 8 (September/October 1984): 3–12.

U.S. Congress. Joint Committee on Taxation. *Taxation of Energy and Natural Resources: Scheduled for Hearings Before the Subcommittee on Energy and Agricultural Taxation of the Committee on Finance on June 21 and 28, 1985.* Washington, DC: GPO, 1985, 35 p.

U.S. Congress. Senate. Committee on Energy and Natural Resources. *Impact of Treasury Department's Tax Reform Proposal on Oil and Gas Industry: Hearing, February 15, 1985.* 99th Congress, 1st session. Washington, DC: GPO, 1985, 343 p.

————. *Proposed Treasury Department's Tax Simplification Plan on the Oil and Gas Industry: Hearing, March 15, 1985.* 99th Congress, 1st session. Washington, DC: GPO, 1985, 228 p.

U.S. Congress. Senate. Committee on Finance. Subcommittee on Energy and Agricultural Taxation. *Tax Incentives to Boost Energy Exploration: Hearing, August 3, 1989, on S. 828.* 101st Congress, 1st session. Washington, DC: GPO, 1990, 72 p.

Walczak, Lee, and others. **"Is a Tax Hike Coming? It Seems Inevitable; The Only Questions Now Are What Kind and When."** *Business Week* (February 3, 1986): 48–53.

Weinstein, Bernard L. **"Energy, Taxes and Growth."** *Society* 22 (November/December 1984): 41–47.

Energy Resources

General

Albrecht, Don E., and others. **"The Impacts of Energy-Resource Projects on Rural Communities in the Western United States,"** in Fear, Frank A., and Harry K. Schwarzweller, eds. *Research in Rural Sociology and Development, 1985.* Greenwich, CT: JAI Press, 1985, 109–123.

Bradshaw, Ted K. **"Power from Communities: The Growth of Municipal Energy Production in California."** *Public Affairs Report* 26 (April/June 1985): 1–16.

Fried, Edward R. **"World Oil Markets: New Benefits, Old Concerns."** *Brookings Review* 4 (Summer 1986): 32–38.

Gowdy, J. M., and J. L. Miller. **"Energy Use in the U.S. Service Sector—An Input-Output Analysis."** *Energy* 12 (May 1987): 555–562.

Hamilton, Kirk. **"Energy Intensiveness and Economic Performance Since 1971."** *Canadian Economic Observer* 1 (December 1988): 4.1–4.18.

Hillbruner, M. **"Energy Source Outlook: Will Supplies Keep Up With Demand?"** *Consulting-Specifying Engineer* 3 (May 1988): 46–52.

Holdren, J. P. **"Energy in Transition."** *Scientific American* 263 (September 1990): 156–163.

Institute of Gas Technology. *IGT World Reserves Survey, 1984: A Survey of United States and Total World Production, Proved Reserves, and Remaining Recoverable Resources of Fossil Fuels and Uranium as of December 31, 1984.* Chicago: 1986, 276 p.

Jorgenson, Dale W. **"The Role of Energy in Productivity Growth."** *Energy Journal* 5 (July 1984): 11–26.

Mackenzie, James J. **"Planning Beyond Oil."** *Forum for Applied Research and Public Policy* 2 (Fall 1987): 6–18.

Mounfield, Peter. **"Energy for the Future?"** *Standard Chartered Review* (June 1986): 2–8.

Nowotny, Kenneth, and James Peach. **"Changes in Energy Consumption, 1970–1989."** *Journal of Economic Issues* 26 (March 1992): 183–196.

Ray, G. F. **"The Decline of Primary Producer Power."** *National Institute Economic Review* (August 1987): 40–45.

Robinson, J. B. **"An Embarrassment of Riches: Canada's Energy Supply Resources."** *Energy* 12 (May 1987): 379–402.

Schürmann, Heinz. **"World Energy: Challenges for Stability."** *OPEC Bulletin* 17 (February 1986): 11–17.

Sullivan, Gene D. **"Prospects for Energy Supplies."** *Federal Reserve Bank of Atlanta Economic Review* 76 (November/December 1991): 17–25.

U.S. Congress. House. Committee on Government Operations. Environment, Energy, and Natural Resources Subcommittee. *Review of Long-Term World Oil Outlook; Petroleum Product Imports; and Energy-Related Tax Reform Proposals: Hearings, April 1–2, 1985.* 99th Congress, 1st session. Washington, DC: GPO, 1985, 901 p.

————. *Review of Outer Continental Shelf Oil and Gas Leasing: Hearing, October 12, 1984.* 98th Congress, 2nd session. Washington, DC: GPO, 1985. 318 p.

U.S. Congress. House. Committee on Science, Space, and Technology. Subcommittee on Energy Research and Development. *Geosciences: Hearings, July 15–16, 1987.* 100th Congress, 1st session. Washington, DC: GPO, 1988, 596 p.

U.S. Congress. Office of Technology Assessment. *Energy Use and the U.S. Economy.* Washington, DC: GPO, 1990, 65 p.

U.S. Energy Information Administration. Office of Energy Markets and End Use. *International Energy Outlook, 1987: Projections to 2000.* Washington, DC: GPO, 1988, 53 p.

————. *International Energy Outlook, 1989: Projections to 2000.* Washington, DC: GPO, 1989, 60 p.

Williams, R. H. **"A Low Energy Future for the United States."** *Energy* 12 (October/November 1987): 929–944.

Energy Review

"Alaska North Slope Oil Production and Revenue Projections." *Alaska Review of Social and Economic Conditions* 22 (February 1985): 1–11.

California Energy Commission. *The 1985 California Energy Plan.* Sacramento, CA: 1985, 44 p.

Carpenter, G. L., and Stanley J. Keller. *Oil Production and Development in Indiana During 1985.* Bloomington, IN: Indiana Department of Natural Resources. Geological Survey, 1986, 49 p.

Macal, C. M., and others. **"An Integrated Energy Planning Model for Illinois."** *Energy* 12 (December 1987): 1239–1250.

Montana Department of Natural Resources and Conservation. Oil and Gas Conservation Division. *Annual Review for the Year 1983, Relating to Oil and Gas.* Helena, MT: 1984, 69 p.

————. *Annual Review for the Year 1984, Relating to Oil and Gas.* Helena, MT: 1985, 82 p.

New York (State) Energy Office. Division of Policy Analysis and Planning. *New York State Annual Review, 1970–1985: Energy Consumption, Supply and Price Statistics, 1970–1985.* Albany, NY: 1986, 97 p.

"State Searches for Energy Solutions." *Journal of State Government* 63 (October/December 1990): 89–102+.

U.S. Energy Information Administration. Office of Energy Markets and End Use. *Annual Energy Outlook, 1986: With Projections to 2000.* Washington, DC: GPO, 1987, 70 p.

————. *Annual Energy Outlook, 1989: With Projections to 2000.* Washington, DC: GPO, 1989, 81 p.

————. *Annual Energy Outlook, 1990: With Projections to 2010.* Washington, DC: 1990, 10–5 p.

————. *Annual Energy Review, 1986.* Washington, DC: GPO, 1987, 293 p.

————. *Annual Energy Review, 1987.* Washington, DC: GPO, 1988, 301 p.

————. *Annual Energy Review, 1988.* Washington, DC: GPO, 1989, 309 p.

————. *International Energy Annual, 1986.* Washington, DC: GPO, 1987, 135 p.

————. *International Energy Annual, 1987.* Washington, DC: GPO, 1988, 141 p.

Cogeneration

Benz, A. D. **"Cogeneration in the Petroleum Refinery."** *Chemical Engineering Progress* 82 (October 1986): 21–27.

Douglas, J. **"The Challenge of Packaged Cogeneration."** *EPRI Journal* 13 (September 1988): 28–37.

Meckler, M. **"Cogeneration: An Up and Coming Technology."** *Consulting-Specifying Engineer* 5 (January 1989): 100–105; 5 (March 1989): 81–85.

Orlando, J. A. **"Packaged Cogeneration for Small Buildings."** *Heating/Piping/Air Conditioning* 60 (December 1988): 49–54.

Peltier, R. V., and J. F. Ring. **"Managing Cogeneration Systems."** *Energy Engineering* 86:5 (1989): 6–25.

Smith, D. J. **"Waste Fuels, Refuse-Derived Fuels and Cogeneration."** *Power Engineering* 90 (April 1986): 40–46.

Smith, D. J., and M. Reynolds. **"Cogeneration, Small Power Production Show Rapid Growth."** *Power Engineering* 90 (October 1986): 27–32.

Smith, D. J., and others. **"Non-Utility Power Generation to Supply 50% of New Capacity."** *Power Engineering* 93 (September 1989): 23–37.

Whipple, D. P., and F. J. Trefny. **"Current Electric System Operating Problems from a Cogenerator's Viewpoint."** *IEEE Transactions on Power Delivery* 4 (August 1989): 1037–1042.

Electric Power

General

Baum, Vladimir. **"The Generation Game: World Electricity Consumption Is Set for Unstoppable Growth."** *Petroleum Economist* 59 (February 1992): 11–15.

Grubb, M. J. **"The Integration of Renewable Electricity Sources."** *Energy Policy* 19 (September 1991): 670–688.

Schweitzer, Martin, and others. **"A Look at the Resource Portfolios of 24 Electric Utilities."** *Electricity Journal* 4 (August/September 1991): 38–45.

U.S. Congress, Senate. Committee on Environment and Public Works. Subcommittee on Nuclear Regulation. *The Role of Nuclear Energy in Meeting Future Electricity Demands: Hearing, August 1, 1990.* 101st Congress, 2nd session. Washington, DC: GPO, 1990, 101 p.

U.S. Laws and Statutes. *Compilation of Selected Energy-Related Legislation: Vol. 2, Electricity.* 100th Congress, 1st session. Washington, DC: GPO, 1987, 215 p.

Market

Bjorklund, Glenn J. **"Planning for Uncertainty at an Electric Utility."** *Public Utilities Fortnightly.* 120 (October 15, 1987): 15–21.

Brown, Ian, ed. **"Energy Efficiency in Electricity."** *Energy Policy* 19 (April 1991): 195–287.

Calzonetti, Frank J., and others. **"U.S. Power Plant Location and Fuel Mix: Opportunities for Electricity Exporting Regions."** *Energy Policy* 14 (December 1986): 528–541.

Glazer, Sarah. **"Deregulating Electric Power: Federal Authorities Want Competitive Bidding Among Power Suppliers."** *Editorial Research Reports* (November 20, 1987): 602–615.

Hayes, Mary Sharpe, and Richard M. Scheer. **"Least-Cost Planning: A National Perspective."** *Public Utilities Fortnightly* 119 (April 2, 1987): 26–33.

Henderson, Yolanda K., and others. **"Planning for New England's Electricity Requirements."** *New England Economic Review* (January/February 1988): 3–30.

Hirst, Eric. **"Electricity: Getting More with Less: The Free Market Is Not Providing Economic Incentives to Use Electricity Efficiently: The Government and Utilities Need to Help."** *Technology Review* 93 (July 1990): 32–38.

Kwaczek, Adrienne S., and William A. Kerr. **"Canadian Exports of Electricity to the U.S.: International Competitiveness or International Risk Bearing?"** *World Competition* 13 (September 1989): 19–28.

Lehr, Ronald L., and Robert Touslee. **"What Are We Bid? Stimulating Electric Generation Resources Through the Auction Method."** *Public Utilities Fortnightly* 120 (November 12, 1987): 11–17.

Martinson, Linda, and Thomas W. Loris. **"The Transitional Bulk Power Market."** *Public Utilities Fortnightly* 120 (November 26, 1987): 19–24.

Mills, Mark P. **"Electricity: Energy for Re-industrializing America."** *Public Utilities Fortnightly* 119 (April 2, 1987): 21–25.

Pierce, Richard J., Jr. **"A Proposal to Deregulate the Market for Bulk Power."** *Virginia Law Review* 72 (October 1986): 1183–1235.

"Power Generation: Rapid Industrialisation and Population Growth Rates Have Placed Unprecedented Demands on Power Supplies in Developing Countries." *South* (May 1988): 85+ 11-page section.

Ramesh, S. **"Peak Load Prediction Using Weather Variables."** *Energy* 13 (August 1988): 671–679.

Schipper, L., and others. **"Residential Electricity Consumption in Industrialized Countries: Changes Since 1973."** *Energy* 12 (December 1987): 1197–1208.

Smartt, Lucien E., comp. **"The Electric Utility Executives Forum."** *Public Utilities Fortnightly* 119 (May 28, 1987): 66–72+.

U.S. Congress, Senate. Committee on Energy and Natural Resources. *U.S. Electricity Supply and Demand: The Northeastern Region: Hearing, April 13, 1989.* 101st Congress, 1st session. Washington, DC: GPO, 1989, 1054 p.

U.S. Energy Information Administration. Office of Coal, Nuclear, Electric and Alternate Fuels. *Annual Outlook for U.S. Electric Power, 1990: Projections Through 2010.* Washington, DC: 1990, 71 p.

Planning

Beneson, Robert. **"Electricity Supply: Surplus or Shortage?"** *Editorial Research Reports* (August 23, 1985): 639–656.

Dotten, Michael C., and Jeffrey A. Boecker. **"Regional Power Planning: The Pacific Northwest Experience."** *Public Utilities Fortnightly* 118 (September 18, 1986): 18–22.

Grunau, John, and Yuwa Wang. **"Dilemma in Electricity Generation Planning."** *Public Utilities Fortnightly* 116 (September 5, 1985): 15–20.

Henderson, Yolanda K., and others. **"Planning for New England's Electricity Requirements."** *New England Economic Review* (January/February 1988): 3–30.

Hirst, Eric, and Martin Schweitzer. **"Uncertainty: A Critical Element of Integrated Resource Planning."** *Electricity Journal* 2 (July 1989): 16–26+.

Mills, Mark P. **"A Return of the Age of Oil for the U.S. Electric Supply System?"** *Public Utilities Fortnightly* 117 (February 6, 1986): 33–40.

Technology

Blair, Peter D. **"New Electric Power Technologies: Cost and Performance in the 1990s."** *Energy Systems and Policy* 10:3 (1987): 189–217.

Bradshaw, Ted K. **"Power from Communities: The Growth of Municipal Energy Production in California."** *Public Affairs Report* 26 (April/June 1985) 1–16.

Clemmensen, Jane M., and Ralph J. Ferraro. **"The Emerging Problem of Electric Power Quality."** *Public Utilities Fortnightly* 116 (November 28, 1985): 18–25.

Hinman, George W., and Thomas C. Lowinger. **"A Comparative Study of Japan and United States Nuclear Enterprise: Industry Structure and Construction Experience."** *Energy Systems and Policy* 11:3 (1987): 205–229.

International Energy Agency. *Electricity End-Use Efficiency.* Washington, DC: Organization for Economic Cooperation and Development, 1989, 200 p. ISBN 92-64-13259-7.

Lindgren, Nilo A. **"Consortium-Style Research and Development Proves Its Worth."** *Public Utilities Fortnightly* 119 (April 2, 1987): 26–33.

Lovins, Amory B. **"The Megawatt Revolution: Using Existing Technology, Says This Expert, We Can Save Three Fourths of All Electricity Used Today."** *Across the Board* 27 (September 1990): 18–23.

McGowan, Francis. **"New Developments in Superconducting Materials: Implications for the Electricity Supply Industry."** *Energy Policy* 16 (October 1988): 506–513.

Mills, Mark P. **"A Return of the Age of Oil for the U.S. Electric Supply System?"** *Public Utilities Fortnightly* 117 (February 6, 1986): 33–40.

New Jersey. General Assembly Energy and Natural Resources Committee. *Public Hearing to Discuss Issues Concerning Impediments to the Growth of Cogeneration in New Jersey: Newark, New Jersey, April 13, 1987.* Trenton, NJ: 1987, 210 p.

"New Push for Energy Efficiency." *EPRI Journal* 15 (April/May 1990): 4–17.

Norgaard, Richard, and others. **"Electric Generation Capacity in New England in 1995."** *Northeast Journal of Business and Economics* 13 (Spring/Summer 1987): 31–46.

Spencer, Dwain F., and Peter H. Benziger. **"Cool Gasification Combined-Cycle Power Plants: A Versatile Clean Coal Technology for the 1990s."** *Public Utilities Fortnightly* 118 (July 24, 1986): 16–23.

Starr, Chauncey, and Milton F. Searl. **"Global Energy and Electricity Futures: Demand and Supply Alternatives."** *Energy Systems and Policy* 14 (January/March 1990): 53–83.

"Transmission: A Special Report." *Public Utilities Fortnightly* 126 (July 19, 1990): 12–31+.

U.S. Congress. House. Committee on Energy and Commerce. Subcommittee on Energy and Power. *Electricity Transmission Access: Hearing, May 24, 1989.* 101st Congress, 1st session. Washington, DC: GPO, 1989, 227 p.

U.S. Congress. House. Committee on Science and Technology. Subcommittee on Energy Development and Applications. *Clean Coal Technologies Initiative: Hearing, May 8, 1985.* 99th Congress, 1st session. Washington, DC: GPO, 1985, 466 p.

U.S. Congress. Senate. Committee on Environment and Public Works. Subcommittee on Environmental Protection. *Responding to the Problem of Global Warming: Hearing, August 10, 1989.* 101st Congress, 1st session. Washington, DC: GPO, 1989, 122 p.

U.S. Energy Information Administration. Office of Coal, Nuclear, Electric, and Alternative Fuels. *Electric Power Annual, 1988.* Washington, DC: 1989, 107 p.

Laws and Regulations

Gunn, Elizabeth M., and others. **"The Public Utility Regulatory Policies Act: Issues in Federal and State Implementation."** *Policy Studies Journal* 13 (December 1984): 353–363.

Hirst, Eric. **"Integrated Resource Planning: The Role of Regulatory Commissions."** *Public Utilities Fortnightly* 122 (September 15, 1988): 33–40+.

Lacy, Robert L. **"Guidelines for Regulators in Evaluating Electric Demand Forecasts."** *Public Utilities Fortnightly* 116 (November 28, 1985): 36–41.

Patrizia, Charles A., and others. **"Regulatory Conflicts Facing Electric Utilities Under the Clean Air Act Amendments of 1990."** *Electricity Journal* 5 (March 1992): 28–36.

U.S. Congress. Senate. Committee on Energy and Natural Resources. Subcommittee on Natural Resources Development and Production. *Domestic Coal Industry: Hearing, May 1, 1986, on the Impact of Coal and Electricity Imports on the Domestic Coal Industry.* 99th Congress, 2nd session. Washington, DC: GPO, 1986, 216 p.

U.S. Laws and Statutes. *Compilation of Selected Energy-Related Legislation: Vol. 2, Electricity.* 100th Congress, 1st session. Washington, DC: GPO, 1987, 215 p.

Whittaker, M. Curtis. **"The Federal Power Act and Hydropower Development: Rediscovering State Regulatory Powers and Responsibilities."** *Harvard Environmental Law Review* 10:1 (1986): 135–187.

Conservation

Hartman, Raymond S. **"Energy Conservation Programmes: The Analysis and Measurement of Their Effects."** *Energy Policy* 14 (October 1986): 413–424.

Hirst, Eric. **"Electric Utility Energy Conservation and Load Management Programmes: R & D Opportunities."** *Energy Policy* 15 (April 1987): 103–108.

Hirst, Eric, and Kenneth Keating. **"Dynamics of Energy Savings Due to Conservation Programs."** *Energy Systems and Policy* 10:3 (1987): 257–273.

Katz, Myron B. **"Utility Conservation Incentives: Everyone Wins: By Allowing Utilities to Charge Consumers for Energy Conservation Services in the Same Way They Are Charged for Energy Supply, Regulators Can Help Serve Both Efficiency and Equity."** *Electricity Journal* 2 (October 1989): 26–35.

McKinsey, Lauren. **"Conserving During Surplus: Montana's Energy Dilemma: Results of a Survey Sponsored by the Northwest Power Planning Council."** *Montana Business Quarterly* 23 (Winter 1985): 18–22.

Nichols, David, and Paul D. Raskin. **"Conservation Utilities: New Force on the Demand Side."** *Electricity Journal* 2 (October 1989): 18–25.

Norland, Douglas, and James L. Wolf. **"Utility Conservation Programs: Opportunities and Strategies."** *Public Utilities Fortnightly* 116 (August 8, 1985): 27–35.

U.S. Congress. House. Committee on Energy and Commerce. Subcommittee on Energy Conservation and Power. *Need for New Powerplants: Hearing, February 7, 1984, on Congressional Research Service and Department of Energy Electric Utility Reports.* 98th Congress, 2nd session. Washington, DC: GPO, 1984, 638 p.

U.S. Congress. Senate. Committee on Energy and Natural Resources. *Clean Coal Technology Deployment Act: Hearing, May 18, 1987, on S. 879, to Encourage the Deployment of Clean Coal Technologies so as to Assure the Development of Additional Electric Generation and Industrial Energy Capacity.* 100th Congress, 1st session. Washington, DC: GPO, 1988, 637 p.

Economic

Bland, F. Paul. **"Problems of Price and Transportation: Two Proposals to Encourage Competition from Alternative Energy Sources."** *Harvard Environmental Law Review* 10:2 (1986): 345–416.

Cavanagh, Ralph C. **"Least-Cost Planning Imperatives for Electric Utilities and Their Regulators."** *Harvard Environmental Law Review* 10:2 (1986): 299–344.

Cofala, J., and others. **"Energy-Economy Modeling: A Survey."** *Energy* 15 (March/April 1990): 387–394.

Davis, G. R. **"Energy for Planet Earth."** *Scientific American* 263 (September 1990): 54–60+.

Devine, Michael D., and others. **"PURPA 210 Avoided Cost Rates: Economic and Implementation Issues."** *Energy Systems and Policy* 11:2 (1987): 85–101.

Hay, Tina M., and Barbara Gray. **"The National Coal Policy Project: An Interactive Approach to Corporate Social Responsiveness,"** in

Preston, Lee, ed. *Research in Corporate Social Performance and Policy, 1985: A Research Annual.* Greenwich, CT: JAI Press, 1985, 191–212.

Naeve, C. M. **"Which Way to Reliability and Price Stability for Tomorrow's Electric Utilities?"** *Public Utilities Fortnightly* 119 (January 1987): 11–14.

Ottinger, Richard L. **"Getting At the True Cost of Electric Power: Many Jurisdictions Are Deciding That Prudent Utility Regulation Requires Consideration of the Total Cost of Energy Resources, Including All Environmental Costs."** *Electricity Journal* 14 (July 1990): 14–23.

Ross, M., and others. **"Energy Demand and Materials Flows in the Economy."** *Energy* 12 (October/November 1987): 953–967.

Spangler, Miller B. **"An International Perspective on Risk and Equity Issues Associated with the Coal and Nuclear Fuel Options."** *Journal of Public and International Affairs* 5 (Winter 1984): 101–121.

Reserves

Dale, Larry L. **"The Pace of Mineral Depletion in the United States [Relation Between the Pace of Copper, Oil, and Coal Depletion and Extraction Cost and Royalty Costs: Some Emphasis on Anthracite Coal]."** *Land Economics* 60 (August 1984): 255–267.

U.S. Congress. House. Committee on Government Operations. Environment, Energy and Natural Resources Subcommittee. *Review of Long-Term World Oil Outlook: Petroleum Product Imports; and Energy-Related Tax Reform Proposals: Hearings, April 1–2, 1985.* 99th Congress, 1st session. Washington, DC: GPO, 1985, 901 p.

U.S. Congress. Senate. Committee on Energy and Natural Resources. Subcommittee on Mineral Resources Development and Production. *Coal Reserves: Hearing, September 18, 1987, on the National Coal Council's Reserve Data Base Report and the State of Information Relating to the Quality and Recoverability of U.S. Coal Reserve.* 100th Congress, 1st session. Washington, DC: GPO, 1988, 98 p.

U.S. Energy Information Administration. Office of Coal, Nuclear, Electric, and Alternate Fuels. *Estimation of U.S. Coal Reserves by Coal Type: Heat and Sulfur Content.* Washington, DC: 1989, 57 p.

Conservation

General

Barrett, Larry B. **"Collaborative Process in Strategic Energy Planning."** *Strategic Planning for Energy and the Environment* 10 (Fall 1990): 35–40.

Brown, S. P. A., and Keith R. Phillips. **"U.S. Oil Demand and Conservation."** *Contemporary Policy Issues* 9 (January 1991): 67–72.

Chen, M. **"Open Protocols for Energy Management Systems: What's Right for You?"** *Energy Engineering* 85:4 (1988): 8–12.

Coltrane, Scott, and others. **"The Social-Psychological Foundations of Successful Energy Conservation Programmes."** *Energy Policy* 14 (April 1986): 133–148.

"Energy Conservation: Tables and Charts." *Forum for Applied Research and Public Policy* 2 (Spring 1987): 64–97.

Gibbons, J. H., and others. **"Strategies for Energy Use."** *Scientific American* 261 (September 1989): 136–143.

Greenhalgh, Geoffrey. **"Energy Conservation Policies."** *Energy Policy* 18 (April 1990): 293–299.

Johansson, T. B., and R. H. Williams. **"Energy Conservation in the Global Context."** *Energy* 12 (October/November 1987): 907–919.

Kenney, W. F. **"Strategies for Conserving Energy."** *Chemical Engineering Progress* 84 (March 1988): 43–48.

Kornbluh, Hy, and others. **"Worker Participation in Energy and Natural Resources Conservation."** *International Labour Review* 124 (November/December 1985): 737–754.

Kushler, Martin G. **"Use of Evaluation to Improve Energy Conservation Programs: A Review and Case Study."** *Journal of Social Issues* 45 (Spring 1989): 153–168.

Morell, Jonathan A., ed. **"Special Feature: Energy Conservation Program Evaluation."** *Evaluation and Program Planning* 12 (1989): 113–206.

Morse, W. L., and H. G. Peach. **"Control Concepts in Conservation Supply."** *Energy* 14 (November 1989): 727–735.

Nichols, David, and Paul D. Raskin. **"Conservation Utilities: New Force on the Demand Side."** *Electricity Journal* 2 (October 1989): 18–25.

Niemeyer, Shirley M., and Earl W. Morris. **"Economic and Psychological Constraints in Household Energy Conservation."** *Housing and Society* 13:1 (1986): 44–55.

Olchefske, Thomas. **"An Emerging Opportunity for Public-Private Partnerships in Energy Conservation."** *Privatization Review* 6 (Winter 1991): 7–15.

Raab, Jonathan. *Waste Not, Want Not: Internal Energy Management in Two Oregon Cities [Corvallis and Salem].* BGRS No. 84–18. Eugene, OR: University of Oregon, Bureau of Governmental Research and Service, 1984, 30 p.

Reisner, Marc. **"The Rise and Fall and Rise of Energy Conservation: Despite an Oil Glut, Congress Has Passed an Appliance Energy Conservation Act that Could Save $28 Billion by the Year 2000 and Eliminate the Need for Twenty-two Power Plants."** *Amicus Journal* 9 (Spring 1987): 22–31.

Stern, Paul C., and others. **"The Effectiveness of Incentives for Residential Energy Conservation"** *Evaluation Review* 10 (April 1986): 147–176.

Vine, E. L., and J. P. Harris. **"Evaluating Energy and Non-Energy Impacts of Energy Conservation Programs: A Supply Curve Framework of Analysis."** *Energy* 15 (January 1990): 11–21.

Efficiency

Berrie, T. W. **"Policy Issues in Improving Energy Efficiency."** *Energy Policy* 15 (December 1987): 29–33.

Brown, Ian, ed. **"Energy Efficiency in Electricity."** *Energy Policy* 19 (April 1991): 195–287.

Fickett, Arnold P., Clark W. Gellings, and Amory B. Lovins. **"Efficient Use of Electricity."** *Scientific American* 263 (September 1990): 64–75.

Hirst, Eric. **"Improving Energy Efficiency in the USA: The Federal Role."** *Energy Policy* 19 (July/August 1991): 567–577.

———. **"Progress and Potential in Evaluating Energy Efficiency Programs."** *Evaluation Review* 14 (April 1990): 192–205.

Kentucky Legislature Research Commission. Program Review and Investigations Committee. *The Weatherization Assistance Program and Other Energy Conservation Efforts in Kentucky.* Frankfort, KY: 1989, 142 p.

Rosenfeld, A. H., and D. Hafemeister. **"Energy-Efficient Buildings."** *Scientific American* 258 (April 1988): 78–85.

Spiller, Larry N. **"The Energy Wolf Is Still Out There."** *Heating/Piping/Air Conditioning* 57 (September 1985): 67–73.

———. **"The Energy Wolf is Still Out There [How Progressive American Companies Are Protecting Themselves Against Another Oil Crisis Through Conservation Planning and Capital Investment]."** *Harvard Business Review* 63 (January/February 1985): 119–127.

Veigel, John M., and Sanford Lakoff. **"U.S. States and Energy Efficiency: North Carolina's 'Quango'."** *Energy Policy* 13 (October 1985): 445–457.

Environment

Baxter, Lester W., and Kevin Calandri. **"Global Warming and Electricity Demand: A Study of California."** *Energy Policy* 20 (March 1992): 233–244.

Brieger, Heidi E. **"LUST and the Common Law: A Marriage of Necessity."** *Boston College Environmental Affairs Law Review* 13:4 (1986): 521–551.

Davis, Joseph A. **"Alaskan Wildlife Refuge Becomes Battleground: Energy Industry vs. Environmentalists."** *Congressional Quarterly Weekly Report* 45 (August 22, 1987): 1939–1943.

Fraas, Arthur, and Albert McGartland. **"Alternative Fuels for Pollution Control: An Empirical Evaluation of Benefits and Costs."** *Contemporary Policy Issues* 8 (January 1990): 62–74.

Fri, Robert W. **"Energy and Environment: A Coming Collision?"** *Resources* (Winter 1990): 1–4.

Gray, C. Boyden. **"Octane, Ozone, and Obstinacy."** *Regulation (Regulation Found)* 11:2 (1987): 37–43.

Hrubovcak, James. **"Ethanol in Agriculture and the Environment."** *Food Review* 14 (April/June 1991): 15–20.

Leavenworth, Geoffrey. **"L.U.S.T.: Leaking Underground Storage Tanks: New Understanding About Them Has Helped Improve Construction Standards, but May Not Forestall a Wave of Claims and Litigation."** *Insurance Review* 42 (March 1987): 20–23+.

Loukissas, Philippos J. **"Impacts of Energy Development on Coastal Recreation: The Case of New Jersey."** *Coastal Zone Management Journal* 13:3/4 (1986): 281–307.

Lovins, A. B. **"Abating Air Pollution at Negative Cost Via Energy Efficiency."** *JAPCA* 39 (November 1989): 1432–1435.

Mackenzie, James J., and Mohamed. T. El-Ashry. **"Ill Winds: Air Pollution's Toll on Trees and Crops."** *Technology Review* 92 (April 1989): 64–71.

Skea, Jim, ed. **"Climate Change: Policy Implications."** *Energy Policy* 19 (March 1990): 90–160.

"Special Report: Troubled Waters." *Amicus Journal* 11 (Summer 1989): 10–31.

U.S. Congress. House. Committee on Merchant Marine and Fisheries. Subcommittee on Fisheries and Wildlife, Conservation and the Environment. *Arctic National Wildlife Refuge: Hearings, Pt. 3, February 24–March 31, 1988.* 100th Congress, 2nd session. Washington, DC: GPO, 1988, 1001 p.

U.S. Congress. Senate. Committee on Commerce, Science, and Transportation. Subcommittee on the Consumer. *Global Warming and CAFE [Corporate Average Fuel Economy] Standards: Hearing, May 2, 1989.* 101st Congress, 1st session. Washington, DC: GPO, 1989, 343 p.

U.S. Congress. Senate. Committee on Energy and Natural Resources. *Arctic National Wildlife Refuge, Alaska: Hearings, Pts. 1–2, June 2, 1987, on the Report of the Secretary of the Interior to the Congress Regarding Oil and Gas Leasing on the Coastal Plain of the Arctic National Wildlife Refuge, Alaska.* 100th Congress, 1st session. Washington, DC: GPO, 1987, 2 parts.

Utton, Albert E., and Paul D. McHugh. **"On an Institutional Arrangement for Developing Oil and Gas in the Gulf of Mexico."** *Natural Resources Journal* 26 (Fall 1986): 717–732.

Veigel, John M., and Sanford Lakoff. "U.S. States and Energy Efficiency: North Carolina's 'Quango'." *Energy Policy* 13 (October 1985): 445–457.

White, David C., and others. "The New Team: Electricity Sources Without Carbon Dioxide." *Technology Review* 95 (January 1992): 42–50.

Technology

Adams, S. G. "Saving Energy Through Design/Build: Commercial." *Heating/Piping/Air Conditioning* 60 (May 1988): 65–71.

Bevington, Rick, and Arthur H. Rosenfeld. "Energy for Buildings and Homes." *Scientific American* 263 (September 1990): 76–86.

Buderi, Robert, and Emily T. Smith. "Conservation Power: It Has a New Look That's Igniting an Energy Revolution." *Business Week* (September 16, 1991): 86–92.

DeLucia, M., and G. Manfrida. "Breakdown of the Energy Balance in a Glass Furnace." *Journal of Energy Resources Technology* 112 (June 1990): 124–129.

"Energy Management Systems." *Specifying Engineer* 56 (October 1986): 49–60.

Gallo, W. L. R., and L. F. Milanez. "Choice of a Reference State for Exergetic Analysis." *Energy* 15 (February 1990): 113–121.

"Innovative Ways to Save Energy in New Buildings." *Heating/Piping/Air Conditioning* 58 (May 1986): 71–99.

Jedlicka, A. D. "Improving Conservation Behavior." *Energy Engineering* 82 (August/September 1985): 28–34.

Lazar, I. A. "Electrical Demand Control—How to Attain It." *Consulting-Specifying Engineer* 2 (October 1987): 76–81.

Residential and Industrial

Berry, Linda, and Martin Schweitzer. "Residential Conservation Programmes for the Elderly: Marketing Techniques and Organizational Structures." *Energy Policy* 19 (July/August 1991): 596–605.

Hirst, Eric, and Richard Goeltz. **"Evaluation of Residential Energy Conservation Programs in Minnesota."** *Evaluation Review* 9 (June 1985): 329–347.

Longstreth, M., and M. Topliff. **"Determinants of Energy Savings and Increases After Installing Energy-Conserving Devices."** *Energy* 15 (June 1990): 523–537.

Meckler, M. **"Indoor Air Quality vs. Energy Efficiency: Impact of New Ventilation Standards."** *Consulting-Specifying Engineer* 4 (July 1988): 82–88.

Ross, Marc H., and Daniel Steinmeyer. **"Energy for Industry."** *Scientific American* 263 (September 1990): 88–98.

Schweitzer, Martin. **"Energy Conservation for Low-Income Households: A Study of the Organization and Outcomes of Weatherization Assistance Programs."** *Energy Systems and Policy* 12:2 (1988): 101–117.

Train, K. E., and P. C. Ignelzi. **"The Economic Value of Energy-Saving Investments by Commercial and Industrial Firms."** *Energy* 12 (July 1987): 543–553.

Vine, Edward, and Jeff Harris. **"Implementing Energy Conservation Programs for New Residential and Commercial Buildings."** *Energy Systems and Policy* 13:2 (1989): 115–139.

Wikler, G. A., and J. Cade. **"Conservation Assessment of Commercial Refrigeration."** *Energy Engineering* 87:5 (1990): 49–78.

Wu, R. H., and others. **"Energy Analysis of an Integrated Steel Mill: A Process Modeling Approach."** *Energy* 14 (December 1989): 831–838.

Laws and Regulations

Byrne, John, and Daniel Rich. **"Deregulation and Energy Conservation: A Reappraisal [Energy Consumption and Conservation Related to Fuel Prices]."** *Policy Studies Journal* 13 (December 1984): 331–343.

Lawrence, C. **"The Federal Commitment to Conservation and Renewables."** *EPRI Journal* 10 (December 1985): 42–46.

U.S. Congress. House. Committee on Energy and Commerce. Subcommittee on Energy Conservation and Power. *Appliance Standards: Hearing, September 10, 1986, on H.R. 5465, a Bill to Amend the Energy Policy and*

Conservation Act with Respect to Energy Conservation Standards for Appliances. 99th Congress, 2nd session. Washington, DC: GPO, 1987, 193 p.

U.S. Congress. House. Committee on Energy and Commerce. Subcommittee on Energy and Power. *Reducing Energy Expenditures in Federal Facilities: Hearings, March 8, 1988, on H.R. 4065, a Bill to Amend the National Energy Conservation Policy Act with Respect to the Energy Policy of the United States.* 100th Congress, 2nd session. Washington, DC: GPO, 1989, 92 p.

————. *Automotive Fuel Efficiency: Hearing, July 13, 1989.* 101st Congress, 1st session. Washington, DC: GPO, 1989, 73 p.

————. *State Energy Conservation Programs: Hearing, May 17, 1989, on H.R. 711, a Bill to Amend the Energy Policy and Conservation Act to Increase the Efficiency and Effectiveness of State Energy Conservation Programs Carried Out Pursuant to Such Act, and for Other Purposes.* 101st Congress, 1st session. Washington, DC: GPO, 1989, 115 p.

————. *TVA Conservation Programs: Hearing, June 29, 1989.* 101st Congress, 1st session. Washington, DC: GPO, 1989, 336 p.

U.S. Congress. House. Committee on Energy and Commerce. Subcommittee on Oversight and Investigations. *Fuel Economy Issues: Hearing April 10, 1989.* 101st Congress, 1st session. Washington, DC: GPO, 1989, 225 p.

U.S. Congress. House. *Federal Facilities Energy Conservation Programs: Joint Hearing, July 11, 1990, Before the Subcommittee on Energy and Commerce, and the Subcommittee on Environment, Energy and Natural Resources of the Committee on Government Operations.* 101st Congress, 1st session. Washington, DC: GPO, 1990, 112 p.

U.S. Congress. Office of Technology Assessment. *Energy Efficiency in the Federal Government: Government By Good Example?* Washington, DC: GPO, 1991, 113 p.

U.S. Congress. Senate. Committee on Energy and Natural Resources. *Energy Production and Conservation Act of 1990: Hearing, August 2, 1990, on S. 2923.* 101st Congress, 2nd session. Washington, DC: GPO, 1990, 381 p.

U.S. Congress. Senate. Committee on Energy and Natural Resources. Subcommittee on Energy Regulation and Conservation. *Federal Energy Management Improvement Act: Hearing, July 30, 1987, on S. 138, to Amend the National Energy Conservation Act to Improve the Federal Energy Management Program and for Other Purposes.* 100th Congress, 1st session. Washington, DC: GPO, 1987, 162 p.

U.S. Congress. Senate. Committee on Energy and Natural Resources. Subcommittee on Energy Research and Development. *Energy Efficiency and Renewable Energy Research, Development and Demonstration: Hearing, June 15, 1989, on S. 488 and S. 964.* 101st Congress, 1st session. Washington, DC: GPO, 1989, 191 p.

Coal Industry

General

Mangum, Garth, and MacLeans, Geo-JaJa. **"The Prospects for Utah Coal."** *Utah Economic and Business Review* 46 (August 1986): 1–15.

Polzin, Paul E. **"Montana's Coal Industry Facing an Uncertain Future."** *Montana Business Quarterly* 23 (Summer 1985): 2–16.

————. **"Montana's Troubled Coal Industry: A 1986 Update."** *Montana Business Quarterly* 24 (Autumn 1986): 2–8.

Rittenberg, Libby, and Ernest H. Manual, Jr. **"A Case Study of Decline in Labor Productivity: Underground Coal Mining Industry, 1960–1976."** *Quarterly Journal of Business and Economics* 25 (Winter 1986): 38–55.

Environmental Aspects

"Coal in New Mexico and the West: The Reagan-Era Giveaway and Why It Matters." *Workbook (Southwest Research and Info Center)* 91 (July/September 1984): 87–102.

Desai, Uday. **"Public Participation in Environmental Policy Implementation: Case of the Surface Mining Control and Reclamation Act."** *American Review of Public Administration* 19 (March 1989): 49–65.

Kirk, Victor. **"Bad Days for Black Rock: Fresh Environmental Challenges and a Bitter Labor-Management Dispute Are Threatening Coal's Image as the Most Stable and Reliable Domestic Energy Resource Available."** *National Journal* 21 (August 5, 1989): 1988–1991.

St. Antoine, Sara L. **"This Land Is Your Land: This Land Is Mineland."** *Environmental Forum* 6 (May/June 1989): 33–35.

Consumption

Canto, Victor A., and Charles W. Kadlec. **"The Shape of Energy Markets to Come."** *Public Utilities Fortnightly* 117 (January 9, 1986): 21–28.

Daneke, Gregory A., ed. **"Symposium: Energy Futures and the Changing Nature of Public/Private Interactions."** *Policy Studies Review* 5 (August 1985): 69–152.

Davis, Joseph A. **"Energy Security: How Real the Threat?"** *Congressional Quarterly Weekly Report* 45 (May 30, 1987): 1115–1120.

Gowen, Marcia, and Fred Hitzhusen. *Economics of Wood vs. Natural Gas and Coal Energy in Ohio.* Research Bulletin 1174. Wooster, OH: Ohio Agricultural Research and Development Center, Ohio State University, 1986, 20 p.

Hartnett, James P., and John S. Mead. **"Coal Research: The Energy of New Success."** *Illinois Issues* 10 (November 1984): 22–27.

Kutler, Edward. **"Energy Forecasting: The Troubled Past of Looking to the Future."** *Public Opinion (American Enterprise Institute)* 8 (December 1985/January 1986): 47–53.

National Coal Association. *Coal 2000: A Forecast for U.S. Coal.* Washington, DC: 1986, 60 p.

Renner, Michael G. **"Shaping America's Energy Future."** *World Policy Journal* 4 (Summer 1987): 383–414.

U.S. Congress. House. Committee on Energy and Commerce. Subcommittee on Fossil and Synthetic Fuels. *Barriers to Increased Coal Utilization: Staff Report, May 1985.* 99th Congress, 1st session. Washington, DC: GPO, 1985, 93 p.

U.S. Energy Information Administration. Office of Energy Markets and End Use. *International Energy Outlook, 1990.* Washington, DC: National Energy Information Center, 1990, 56 p.

———. *State Energy Data Report: Consumption Estimates, 1960–1987.* Washington, DC: 1989, 468 p.

Technology

Allen, William H. **"Coal Research Aimed at No. 1 Enemy: Sulfur."** *Illinois Issues* 12 (April 1986): 15–18.

Moore, James R. **"Secondary Coal Recovery: A New and Struggling Industry."** *Illinois Business Review* 43 (December 1986): 8–10.

U.S. Congress. House. Committee on Banking, Finance and Urban Affairs. Subcommittee on Economic Stabilization. *Impact of Proposed Clean Air Legislation on Energy Security and Economic Stability: Hearing, March 28, 1990.* 101st Congress, 2nd session. Washington, DC: GPO, 1990, 607 p.

U.S. Congress. House. Committee on Science, Space and Technology. *Global Prospects for U.S. Coal and Coal Technologies: Hearings, July 21 and 23, 1987, Before the Subcommittee on Energy Research and Development and the Subcommittee on International Scientific Cooperation.* 100th Congress, 1st session. Washington, DC: GPO, 1987, 265 p.

U.S. Congress. House. Committee on Science and Technology. Subcommittee on Energy Development and Applications. *Clean Coal Technologies Initiative: Hearing, May 8, 1985.* 99th Congress, 1st session. Washington, DC: GPO, 1985, 466 p.

U.S. Congress. Senate. Committee on Energy and Natural Resources. Subcommittee on Energy Research and Development. *Department of Energy's Fossil Energy Research and Development and Clean Coal Technology Programs: Hearing, July 11, 1989.* 101st Congress, 1st session. Washington, DC: GPO, 1989, 254 p.

U.S. Congress. Senate. Committee on Energy and Natural Resources. Subcommittee on Mineral Resources Development and Production. *Availability, Production, and Distribution of Clean-Burning Fuels: Hearing, May 25, 1989.* 101st Congress, 1st session. Washington, DC: GPO, 1989, 380 p.

Strip Mining

Dernbach, John C. **"Pennsylvania's Implementation of the Surface Mining Control and Reclamation Act: An Assessment of How 'Cooperative Federalism' Can Make State Regulatory Programs More Effective."** *University of Michigan Journal of Law Reform* 19 (Summer 1986): 903–967.

Harris, Richard A. **"Business Responses to Surface Mining Regulation,"** in Preston, Lee E., ed. *Research in Corporate Social Performance and Policy, 1985.* Greenwich, CT: JAI Press, 1985, 73–101.

Transportation

Barber, G. M. **"Regional Transport Investment Planning."** *Annals of the Association of American Geographers* 68:3 (September 1978): 384–395.

King, L., E. Casetti, J. Odland, and K. Semple. **"Optimal Transportation Patterns of Coal in the Great Lakes Region."** *Economic Geography* 47 (July 1971): 401–413.

Exports-Imports

Ballert, A. **"Great Lakes Coal Trade: Present and Future."** *Economic Geography* 29 (January 1953): 48–59.

Kuby, Michael, Samuel Ratick, and Jeffrey Osleeb. **"Modeling U.S. Coal Export Planning Decisions."** *Annals of the Association of American Geographers* 81:4 (December 1991): 627–647.

Schulz, Walter. **"The Supply Potential of the International Steam Coal Market."** *Columbia Journal of World Business* 19 (Spring 1984): 62–69.

U.S. Congress. Senate. Committee on Energy and Natural Resources. Subcommittee on Natural Resources Development and Production. *Impact of Coal Imports on the Domestic Coal Industry: Hearing, June 6, 1985.* 99th Congress, 1st session. Washington, DC: GPO, 1985, 268 p.

————. *Domestic Coal Industry: Hearing, May 1, 1986, on the Impact of Coal and Electricity Imports on the Domestic Coal Industry.* 99th Congress, 2nd session. Washington, DC: GPO, 1986, 216 p.

————. *Prospects of Exporting American Coal: Hearing, August 5, 1986.* 99th Congress, 2nd session. Washington, DC: GPO, 1987, 214 p.

Laws and Regulations

Stanfield, Rochelle L. **"Things Should Be Looking Up for Coal Industry, But Its Future Is Clouded: Production and Consumption Will Hit Record Highs This Year, But the Industry Says Federal Policies [on Mining Safety and Environmental Issues] Have Reduced Coal's Competitiveness with Other Energy Sources."** *National Journal* 16 (August 11, 1984): 1525–1529.

Trebing, Harry M. **"Apologetics of Deregulation in Energy and Telecommunications: An Institutionalist Assessment."** *Journal of Economic Issues* 20 (September 1986): 613–632.

U.S. Congress. House. Committee on Interior and Insular Affairs. *Coal Pipeline Act of 1987: Hearing, September 22, 1987, on H.R. 1531 to Amend the Mineral Lands Leasing Act of 1920 with Respect to the Movement of Coal Over Public Lands; and for Other Purposes.* 100th Congress, 1st session. Washington, DC: GPO, 1988, 251 p.

——. *Coal Pipeline Act of 1989: Hearing, April 11, 1989, on H. 402 to Amend the Minerals Leasing Act of 1920 with Respect to the Movement of Coal Over Public Funds, and for Other Purposes.* 101st Congress, 1st session. Washington, DC: GPO, 1989, 343 p.

U.S. Congress. Senate. Committee on Energy and Natural Resources. *Coal Distribution and Utilization Act of 1987: Hearing, September 10, 1987, on S. 801 to Facilitate the National Distribution and Utilization of Coal.* 100th Congress, 1st session. Washington, DC: GPO, 1988, 320 p.

U.S. Congress. Senate. Committee on Energy and Natural Resources. Subcommittee on Energy and Mineral Resources. *Condition of America's Coal Industry: Hearing, November 15, 1983.* 98th Congress, 1st session. Washington, DC: GPO, 1984, 439 p.

——. *Coal Use By the Nation's Railroads: Hearings, November 28, 1983–January 19, 1984, on the Reintroduction of Coal as Fuel for Railroad Locomotives.* 98th Congress, 1st and 2nd sessions. Washington, DC: GPO 1984, 645 p.

U.S. Congress. Senate. Committee on Energy and Natural Resources. Subcommittee on Natural Resources Development and Production. *State of the Coal Industry in Oklahoma: Hearing, June 14, 1985, on the State of the Coal Industry in Oklahoma, with Emphasis on the Impact of Federal Regulations.* 99th Congress, 1st session. Washington, DC: GPO, 1985, 214 p.

Petroleum Industry

General

Bolze, Dorene A. **"Outer Continental Shelf Oil and Gas Development in the Alaskan Arctic."** *Natural Resources Journal* 30 (Winter 1990): 17–64.

Bolze, Dorene A., and M. B. Lee. **"Offshore Oil and Gas Development: Implications for Wildlife in Alaska."** *Marine Policy* 13 (July 1989): 231–248.

Broadman, Harry G. **"The Social Cost of Imported Oil."** *Energy Policy* 14 (June 1986): 242–252.

Fri, Robert W. **"Direct Action on Oil: We Have a Variety of Options, Both Immediate and Long-Term, for Reducing Dependence on OPEC."** *Issues in Science and Technology* 5 (Spring 1989): 82–87.

Fulkerson, William, Roddie R. Judkins, and Manorj K. Sanghvi. **"Energy from Fossil Fuels."** *Scientific American* 263 (September 1990): 128–135.

Gould, Gregory. *OCS National Compendium: Outer Continental Shelf Oil and Gas Information Through September 1988*. Herndon, VA: U.S. Department of the Interior, Minerals Management Service, OCS Information Program, 1989, 131 p.

Illinois. Department of Energy and Natural Resources. State Geological Survey Division. *Petroleum Industry in Illinois, 1983*. Illinois Petroleum 126. Champaign, IL: 1985, 140 p.

Jones, G. Kevin. **"Major Issues in Developing Alaska's Outer Continental Shelf Oil and Gas Resources."** *Alaska Law Review* 1 (Winter 1984): 209–275.

Kahley, William J., and Gustavo A. Uceda. **"Louisiana: The Worst May Be Over."** *Federal Reserve Bank of Atlanta Economic Review* (November/December 1986): 47–60.

Perlo, Victor. **"The New Crisis in Oil."** *Political Affairs* 65 (June 1986): 10–16.

U.S. Congress. Senate. Committee on Energy and Natural Resources. *Domestic Petroleum Industry Outlook: Hearings, February 9–March 2, 1987*. 100th Congress, 1st session. Washington, DC: GPO, 1987, 664 p.

Utton, Albert E., and Paul D. McHugh. **"On an Institutional Arrangement for Developing Oil and Gas in the Gulf of Mexico."** *Natural Resources Journal* 26 (Fall 1986): 717–732.

"World Oil: Options and Alternatives." *Congressional Research Service Review* 9 (March 1988): 1–23.

Technology

Campbell, Luther L. **"Restructuring for Higher Yields in the Petroleum Industry: To Enhance Their Market Value, a Growing Number of Oil and Gas Companies Are Turning to Creative Forms of Financial and Structural Reorganization."** *Price Waterhouse Review* 30:2 (1986): 32–40.

Matthews, Downs. **"Producing Oil on Public Lands: The Federal Government Owns the Nation's Outer Continental Shelf, Plus One-third of the Total Land Area of the U.S., Estimated to Contain More Than Half of the Nation's Undiscovered Energy Resources."** *Lamp* 70 (Spring 1988): 6–11.

"Relocating the Navajo: When Technology and Tradition Clash." *Technology Review* 89 (July 1986): 46–57+.

U.S. Congress. House. Committee on Small Business. Subcommittee on Energy and Agriculture. *Underground Storage Tanks: Hearing, November 18, 1987.* 100th Congress, 1st session. Washington, DC: GPO, 1988, 324 p.

Walker, William. **"Information Technology and the Use of Energy."** *Energy Policy* 13 (October 1985): 458–476.

Consumption

Cutler, Tom. **"The Military Demand for Oil."** *Petroleum Economist* 52 (August 1985): 279–284; 53 (April 1986): 137–142; 53 (October 1986): 370–373; 54 (May 1987); 161–164; 54 (June 1987): 221–225.

Dunn, John R., and E. Eldon Eversull. **"Cooperatives Supply Nearly 40 Percent of Petroleum Products to Agriculture."** *Farmer Cooperatives* 57 (June 1990): 4–6.

Roeber, Joe, and Alan Kennington. **"Futures Trading and the Oil Industry: A Slow Revolution."** *Journal of Energy and Natural Resources Law* 3:1 (1985): 21–30.

U.S. Congress. House. Committee on Energy and Commerce. Subcommittee on Energy and Power. *Energy: Free Trade with Canada: Hearings, March 1, 1988, Oil and Natural Gas; March 9, 1988, Electricity and Uranium.* 100th Congress, 2nd session. Washington, DC: GPO, 1988, 286 p.

Exports-Imports

U.S. Congress. Senate. Committee on Energy and Natural Resources. Subcommittee on Energy Regulation and Conservation. *Residential Energy Efficiency Ratings Act: Hearing, March 27, 1990, on S. 1355, to Assist Private Industry in Establishing a Uniform Residential Energy Efficiency Rating System, and for Other Purposes.* 101st Congress, 2nd session. Washington, DC: GPO, 1990, 223 p.

"World Oil Markets and U.S.-Arab Relations." *American-Arab Affairs* (Summer 1985): 56–103.

Laws and Regulations

Egghart, Heinrich K. **"The Combined Hydrocarbon Leasing Act of 1981 and Hydrocarbon Extraction in Utah's Tar Sand Triangle."** *Journal of Energy Law and Policy* 8:1 (1987): 119–150.

McDonald, Stephen L. **"An Oil Tariff for National Security? Is a 'Free Market Solution' an Appropriate One from a National Point of View?"** *Review of Regional Economics and Business* 11 (October 1986): 3–9.

Stanfield, Rochelle L. **"Going It Alone: Some Independent Oil Firms Want the Government's Help in Coping with the Oil Price Bust; But Nonindustry Analysts Say Uncle Sam Has Protected the Industry Far Too Long."** *National Journal* 18 (March 29, 1986): 752–758.

U.S. Congress. Joint Committee on Taxation. *Taxation of Petroleum Imports: Scheduled for Hearings on February 27–28, 1986, Before the Subcommittee on Energy and Agricultural Taxation of the Senate Committee.* Washington, DC: GPO, 1986, 32 p.

U.S. Congress. Office of Technology Assessment. *U.S. Oil Production: The Effect of Low Oil Prices.* Washington, DC: GPO, 1987, 24 p.

U.S. Congress. Senate. Committee on Energy and Natural Resources. *Outer Continental Shelf Oil and Gas Moratoria: Hearing, September 17, 1985, on the Impact of Moratoria on OCS Leasing in Federal Waters Adjacent to the Coastline of the State of California.* 99th Congress, 1st session. Washington, DC: GPO, 1986, 703 p.

————. *Domestic and International Petroleum Situation and the Implications of Fees on Imported Oil: Hearing, September 16, 1986.* 99th Congress, 2nd session. Washington, DC: 1987, 211 p.

————. *Domestic Petroleum Industry Outlook: Hearings, February 9–March 2, 1987.* 100th Congress, 1st session. Washington, DC: GPO, 1987, 664 p.

————. *Energy Security: Hearing, May 20, 1987.* 100th Congress, 1st session. Washington, DC: GPO, 1987, 195 p.

————. *Arctic Coastal Plain Public Lands Leasing Act of 1987: Hearings, October 13–22, 1987, on S. 1217, to Amend the Mineral Leasing Act of 1920 to Authorize the Secretary of the Interior to Lease, in an Expeditious and Environmentally Sound Manner, the Public Lands within the Coastal Plain of the North Slope of Alaska for Oil and Gas Exploration, Development, and Production.* 100th Congress, 1st session. Washington, DC: GPO, 1988, 1334 p.

————. *Arctic Coastal Plain Competitive Oil and Gas Leasing Act: Hearing, March 6, 1989, on S. 406, to Authorize Competitive Oil and Gas Leasing on the Coastal Plain of the Arctic National Wildlife Refuge in a Manner Consistent with Protection of the Environment and for Other Purposes.* 101st Congress, 1st session. Washington, DC: GPO, 1989, 1161 p.

U.S. Congress. Senate. Committee on Energy and Natural Resources. Subcommittee on Mineral Resources Development and Production. *Royalty Management Program: Hearing: Pts. 1–2, May 23 and July 12, 1988, on the Department of the Interior's Royalty Management Program.* 100th Congress, 2nd session. Washington, DC: GPO, 1988, 2 pts.

U.S. Congress. Senate. Committee on Finance. *Impact of Imports and Foreign Investment on National Security: Hearing, March 25, 1987, on S. 285 [and other Bills].* 100th Congress, 1st session. Washington, DC: GPO, 1987, 166 p.

U.S. Congress. Senate. Committee on Finance. Subcommittee on Energy and Agricultural Taxation. *Taxation of Imported Oil: Hearings, February 27–28, 1986.* 99th Congress, 2nd session. Washington, DC: GPO, 1986, 768 p.

Economic

Anders, Gary C. **"Oil, Economic Dependence, and Alaska's Development."** *Journal of Energy and Development* 11 (Spring 1986): 243–261.

Bernard, Jean-Thomas, and Robert J. Werner. *Multinational Corporations, Transfer Prices, and Taxes: Evidence from the U.S. Petroleum Industry.* Cambridge, MA: National Bureau of Economic Research, 1989, 40 p.

"Can We Afford Cheap Oil?" *Washington Quarterly* 9 (Summer 1986): 137–164.

Craine, Steve. **"Behind the Sharp Decline in Oil Prices: Market Shake-up Exposes Vulnerability of OPEC."** *Intercontinental Press* 24 (May 5, 1986): 289–291.

Fried, Neal, and others. **"The Oil Patch Slide: How Alaska's Economy Compares to Other Oil States."** *Alaska Economic Trends* 6 (December 1986): 1–10.

Hausker, Karl. **"Oil Import Fees: Measuring the Costs and Benefits."** *Journal of Energy and Development* 13 (Spring 1988): 171–185.

Ikenberry, G. John. **"Market Solutions for State Problems: The International and Domestic Politics of Americal Oil Decontrol."** *International Organization* 42 (Winter 1988): 151–177.

Parker, Marcia A. **"Industry Profile: Oil and Gas."** *Buyouts and Acquisitions* 4 (July/August 1986): 24–30.

U.S. Congress. House. Committee on Small Business. Subcommittee on Energy, Environment, and Safety Issues Affecting Small Business. *Oil Company Mergers: Impact on Petroleum Retailers: Hearing, April 28, 1984.* 98th Congress, 2nd session. Washington, DC: GPO, 1984, 50 p.

Van Lear, William. **"The Restructuring of the Oil Industry."** *Journal of Economic Issues* 23 (March 1989): 147–154.

Zakariya, Hasan S. **"Insurance Against the Political Risks of Petroleum Investment."** *Journal of Energy and Natural Resources Law* 4:4 (1986): 217–229.

Transportation

Hanelt, R. L., and D. S. Smith. **"The Dynamics of West Coast Container Port Competition."** *Transportation Research Forum* 28 (1987): 82–91.

Hayuth, Yehuda. **"Rationalization and Deconcentration of the U.S. Container Port System."** *Professional Geographer* 40 (August 1988): 279–288.

Kenyon, James. **"Elements in Inter-port Competition in the United States."** *Economic Geography* 46 (January 1970): 1–24.

Pagano, Michael A. **"Old Wine in New Bottles? An Analysis and Preliminary Appraisal of the Surface Transportation Assistance Act of 1982."** *Publius* 16 (Winter 1986): 181–197.

Tarpgaard, Peter T. **"U.S. Shipping Subsidies and Ocean Shipping of Petroleum: Policy Choices."** *Contemporary Policy Issues* 4 (October 1986): 79–92.

Thuong, Le T., and Lisa A. Elvey. **"The End of the Supertanker Era."** *Transportation Journal* 25 (Winter 1985): 4–17.

OPEC

Aburdene, Odeh. **"Energy, Investment and Trade: The U.S. and the Gulf."** *Fletcher Forum* 8 (Summer 1984): 273–283.

Ahrari, Mohammed E. **"OPEC and the Hyperpluralism of the Oil Market in the 1980s."** *International Affairs* 61 (Spring 1985): 263–277.

Al-Sahlawi, Mohammed A. **"The Role of OPEC in the World Oil Market."** *Journal of Energy and Development* 14 (Spring 1989): 213-220.

Baum, Vladimir. **"Third World: Oil's Social and Economic Effects."** *Petroleum Economist* 53 (October 1968): 376–378.

Conant, Melvin A. **"Recognizing U.S.-Arab Interdependence: The U.S. Stake in Gulf Oil."** *American-Arab Affairs* (Spring 1987): 57–61.

"East-West Economic Relationships." *Economic Bulletin for Europe* 38 (October 1986): 21–48.

Fieleke, Norman S. **"The Decline of the Oil Cartel."** *New England Economic Review* (July/August 1986): 32–41.

————. **"Oil Shock III?"** *New England Economic Review* (September/October 1990): 3–10.

Georgiou, George C. **"Oil Market Instability and a New OEEC."** *World Policy* 4 (Spring 1987): 295–312.

Grossack, Irvin M. **"OPEC and the Antitrust Laws."** *Journal of Economic Issues* 20 (September 1986): 725–741.

Hunter, Shireen T. **"The Gulf Economic Crisis and Its Social and Political Consequences."** *Middle East Journal* 40 (Autumn 1986): 593–613.

Licklider, Roy. **"The Power of Oil: The Arab Oil Weapon and the Netherlands, the United Kingdom, Canada, Japan, and the United States."** *International Studies Quarterly* 32 (June 1988): 205–226.

Mead, Walter J. **"The OPEC Cartel Thesis Reexamined: Price Constraints from Oil Substitutes."** *Journal of Energy and Development* 11 (Spring 1986): 213–242.

Miremadi, Assadollah. **"OPEC Countries and the Issue of Climate Change."** *OPEC Bulletin* 22 (July/August 1991): 11–18.

Morse, Edward L. **"After the Fall: The Politics of Oil."** *Foreign Affairs* 64 (Spring 1986): 792–811.

"OPEC Aims to Restore Order to Market." *OPEC Bulletin* 17 (May 1986): 4–8.

Ortez, Rene G. **"The OPEC Role Until the Year 2000."** *Fletcher Forum* 8 (Summer 1984): 285–290.

Perlo, Victor. **"The New Crisis in Oil."** *Political Affairs* 65 (June 1986): 10–16.

"The Post-OPEC Era." *World Press Review* 33 (April 1986): 33–38.

Sadare, R. A. **"Oil Exploration in Hostile Environments: Trends and Prospects."** *OPEC Bulletin* 17 (November 1986): 8–14.

Schuler, G. Henry M. **"A Petroleum Forecast: The Impact on U.S.-Arab Relations in the Coming Years."** *American-Arab Affairs* (Winter 1986/1987): 83–87.

Stevens, Russell. **"Trade Finance in the Gulf."** *Euromoney Trade Finance Report* (April 1987): 29–35.

Takin, Manouchehr. **"Energy Cycles: Can They Be Avoided?"** *OPEC Bulletin* 19 (October 1988): 6–10+.

U.S. Congress. House. Committee on Energy and Commerce. *Energy Impact of the Persian Gulf Crisis: Hearing, January 9, 1991.* 102nd Congress, 1st session. Washington, DC: GPO, 1991, 189 p.

Williams, Harold R., and Randall I. Mount. **"OPEC and the U.S. Demand for Motor Gasoline: Short-Run and Long-Run Price Elasticities."** *Rivista Internazionale de Scienze Economiche e Commorciali* 34 (January/February 1987): 147–157.

Natural Gas Industry

General

Adelman, M. A. **"Western Hemisphere Perspectives: Oil and Natural Gas."** *Contemporary Policy Issues* 3 (Summer 1985): 3–12.

Berry, John M. **"Large Problems Still Loom Over National Energy Policy."** *Financier* 13 (September 1989): 6–10.

"Green and Growing Markets: The Natural Gas Industry Is Well Prepared to Take Advantage of New Opportunities in the Environmental 90s." *Petroleum Economist* 57 (August 1990): 8–11.

Harper, John A., and Cheryl L. Cozart. *Oil and Gas Developments in Pennsylvania in 1987.* Pennsylvania. Department of Environmental Resources. Office of Resources Management. Pennsylvania Geological Survey, 4th session. 1988, 80 p. ISBN 0-8182-0110-X.

Hough, G. Vernon. **"World Natural Gas Survey: Another Year of Good Progress."** *Petroleum Economist* 53 (August 1986): 281–283.

Mueller, Michael J., and others. **"Economic Impacts of Integrated Energy Systems: A Study of Potential Positive Impacts to Oklahoma's Economy of Natural Gas Combined Cycle Power Generating Plants that Also Recover CO$_2$."** *Review of Regional Economics and Business* 10 (April 1985): 3–12.

Nulty, Peter. **"Guess Which Fuel IS Looking Hot: Natural Gas Is Clean, Efficient, and Cheap: Most Important, the U.S. Has Lots of It."** *Fortune* 115 (June 8, 1987): 94–96+.

Schroer, Edmund A. **"The State of Natural Gas, 1986."** *Public Utilities Fortnightly* 118 (October 16, 1986): 11–13.

Skancke, Nancy J. **"The Natural Gas Industry of the Future: The Producer: The Picture as of August 1, 1984."** *Policy Studies Journal* 13 (December 1984): 377–387.

Stewart, Robert W. **"Natural Gas on a Frontier of New Challenges."** *Public Utilities Fortnightly* 119 (May 14, 1987): 9–14.

Swanson, Larry D. **"The Oil and Gas Industry in Montana."** *Montana Business Quarterly* 27 (Spring 1989): 12–17.

Tussing, Arlon R. **"A Perspective on Tomorrow for the Changing Gas Industry."** *Public Utilities Fortnightly* 120 (October 1, 1987): 14–21.

U.S. Congress. Senate. Committee on Energy and Natural Resources. Subcommittee on Energy Regulation. *North American Natural Gas Reserves and Resources: Hearing, April 26, 1984.* 98th Congress, 2nd session. Washington, DC: GPO, 1984, 387 p.

U.S. Congress. Senate. Committee on Energy and Natural Resources. Subcommittee on Mineral Resources Development and Production.

Future of the Domestic Oil and Gas Industry: Hearing, April 11, 1990. 101st Congress, 2nd session. Washington, DC: GPO, 1990, 153 p.

U.S. Energy Information Administration. Office of Oil and Gas. *Gas Supplies of Interstate Natural Gas Pipeline Companies, 1988.* Washington, DC: GPO, 1989, 123 p.

Exports-Imports

Price, Robert S. **"Government Policy and International Natural Gas Trade."** *Journal of Energy and Development* 11 (Spring 1986): 189–211.

Supply and Demand

Chamberlin, John H., and Ed R. Mayberry. **"End-Use Fuel Switching: Is It Fair?"** *Electricity Journal* 4 (October 1991): 38–43.

Toman, Michael A. **"The Outlook for 'Spot' Trade in Natural Gas."** *Public Utilities Fortnightly* 118 (July 10, 1986): 28–32.

U.S. Congress. Senate. Committee on Energy and Natural Resources. Subcommittee on Mineral Resources Development and Production. *Natural Gas Supply and Deliverability: Hearing, September 26, 1989.* 101st Congress, 1st session. Washington, DC: GPO, 1990, 484 p.

U.S. Energy Information Administration. Office of Oil and Gas. *Gas Supplies of Interstate Natural Gas Pipeline Companies, 1989.* Washington, DC: GPO, 1990, 119 p.

Utton, Albert E., and Paul D. McHugh. **"On an Institutional Arrangement for Developing Oil and Gas in the Gulf of Mexico."** *Natural Resources Journal* 26 (Fall 1986): 717–732.

Woods, Thomas J. **"The Outlook for Gas Supply: Surviving the Short Term."** *Journal of Energy and Development* 12 (Autumn 1986): 49–66.

Laws and Regulations

Broadman, Harry G. **"Natural Gas Deregulation: The Need for Further Reform."** *Journal of Policy Analysis and Management* 5 (Spring 1986): 496–916.

Jenkins-Smith, Hank C., ed. **"Natural Gas Regulation in the Western U.S.: Perspectives on Regulation in the Next Decade."** *Natural Resources Journal* 27 (Fall 1987): 771–863.

Lambert, Jeremiah D. **"Bypass in the Natural Gas Industry: The Fruit of Regulatory Change."** *Public Utilities Fortnightly* 117 (April 3, 1986): 11–17.

Steelman, J. D., Jr. **"Deregulation of the Natural Gas Industry: How the Free Market Efficiently Allocates Energy Resources."** *Freeman* 36 (June 1986): 16–21.

U.S. Congress. House. Committee on Energy and Commerce. Subcommittee on Fossil and Synthetic Fuels. *Natural Gas Issues: Hearings, June 4–9, 1986, on H.R. 2734, a Bill to Amend the Natural Gas Act of 1978 to Facilitate the Transition of the Natural Gas Industry to a More Competitive Market and for Other Purposes.* 99th Congress, 2nd session. Washington, DC: GPO, 1987, 504 p.

U.S. Congress. Senate. Committee on Energy and Natural Resources. *Natural Gas Wellhead Decontrol: Hearing, May 8, 1989, on S. 625 [and Other Bills].* 101st Congress, 1st session. Washington, DC: GPO, 1989, 332 p.

U.S. Congress. Senate. Committee on Energy and Natural Resources. Subcommittee on Energy Regulation and Conservation. *Natural Gas Market: Hearings, June 18 and July 11, 1985, the Current State of the Factors Affecting the Natural Gas Market.* 99th Congress, 1st session. Washington, DC: GPO, 1986, 567 p.

————. *Implications of Proposed National Energy Policy Legislation for Natural Gas: Hearing, April 3, 1991, on S. 341 (Title X), to Reduce the Nation's Dependence on Imported Oil, to Provide for the Energy Security of the Nation and for Other Purposes.* 102nd Congress, 1st session. Washington, DC: GPO, 1991, 270 p.

Zycher, Benjamin. *Policy Analytics of Natural Gas Decontrol.* International Institute for Economic Research. Minneapolis, MN: Northstar Printing & Mailing, 1985, 29 p.

Economics

Baly, Michael, III, and Randall V. Griffin. **"The Tax Reform Process as Seen by the Natural Gas Industry."** *Public Utilities Fortnightly* 119 (February 19, 1987): 19–22; (March 5, 1987): 16–20.

Brown, Stephen P. **"Consumers May Not Benefit from Wellhead Price Controls on Natural Gas."** *Federal Reserve Bank of Dallas Economic Review* (July 1985): 1–11.

Gowen, Marcia, and Fred Hitzhusen. *Economics of Wood vs. Natural Gas and Coal Energy in Ohio.* Research Bulletin 1174. Wooster, OH: Ohio Agricultural Research and Development Center, Ohio State University, 1986, 20 p.

Niskanen, William A. **"Natural Gas Price Controls: An Alternative View."** *Regulation (Regulation Found)* 10 (November/December 1986): 46–50.

Percebois, Jacques. **"Gas Market Prospects and Relationships with Oil Prices."** *Energy Policy* 14 (August 1986): 329–346.

U.S. Congress. Senate. Committee on Finance. Subcommittee on Energy and Agricultural Taxation. *Tax Incentives to Boost Energy Exploration: Hearing, August 3, 1989, on S. 828.* 101st Congress, 1st session. Washington, DC: GPO, 1990, 72 p.

Conservation

Cady, William A. **"Marginal Conservation Energy: What It Is and How It Is Measured."** *Public Utilities Fortnightly* 118 (November 27, 1986): 27–31.

U.S. Congress. House. Committee on Energy and Commerce. Subcommittee on Energy and Power. *Natural Gas Price Controls: Hearing, April 5, 1989, on H.R. 1595, a Bill to Amend the Natural Gas Policy Act of 1978, and for Other Purposes.* 101st Congress, 1st session. Washington, DC: GPO, 1989, 184 p.

Renewable Energy

General

"Alternative Energy." *IEE Proceedings Part A* 134 (May 1987): 369–463.

Bland, F. Paul. **"Problems of Price and Transportation: Two Proposals to Encourage Competition from Alternative Energy Sources."** *Harvard Environmental Law Review* 10:2 (1986): 345–416.

Cannon, James S., and others. **"The Feud Over Alternative Fuels: Can New Fuels Reconcile the Environment and the Automobile?"** *Environmental Forum* 6 (September/October 1989): 18–29.

Derrick, Stephen. **"Methanol from Natural Gas."** *National Economic Review* (September 1989): 71–74.

Fenn, Scott A. **"Renewable Power Generation: Beyond the Stakeout."** *Public Utilities Fortnightly* 118 (November 13, 1986): 24–29.

Grinnell, Gerald E. **"Ethanol Fuel: The Policy Issues."** *Forum for Applied Research and Public Policy* 3 (Winter 1988): 48–55.

Grubb, M. J. **"The Integration of Renewable Electricity Sources."** *Energy Policy* 19 (September 1991): 670–688.

Hassanain, Mahjoob A. **"Future Prospect for Alternative Sources of Energy."** *Journal of Energy and Development* 10 (Spring 1985): 231–237.

Holt, Don, ed. **"Renewable Resources."** *National Forum* 68 (Summer 1988): 2–43.

Jackson, Tim, ed. **"Renewable Series—Collection."** *Energy Policy* 19 (October 1991): 706–807.

Jain, B.C. **"Rural Energy Centres Based on Renewables—Case Study on an Effective and Viable Alternative."** *IEEE Transactions on Energy Conversion* 2 (September 1987): 329–335.

Kahn, Edward, and Charles A. Goldman. **"Impact of Tax Reform on Renewable Energy and Cogeneration Projects."** *Energy Economics* 9 (October 1987): 215–226.

Kraft, Michael E., and Regina S. Axelrod. **"Political Constraints on Development of Alternative Energy Sources: Lessons from the Reagan Administration."** *Policy Studies Journal* 13 (December 1984): 319–330.

Lenssen, Nicholas, and John E. Young. **"Filling Up in the Future: Running Cars on Methanol to Reduce Urban Air Pollution Is—at Best—a Misguided Notion, One We Can Live Without: Better Transport Options Are on the Way."** *World Watch* 3 (May/June 1990): 18–26.

Mackenzie, James J. **"Planning Beyond Oil."** *Forum for Applied Research and Public Policy* 2 (Fall 1987): 6–18.

McGowan, Francis. **"Controlling the Greenhouse Effect: The Role of Renewables."** *Energy Policy* 19 (March 1991): 110–118.

Mideke, M. **"Alternative Energy—An Overview of Options and Requirements."** *QST* 71 (September 1987): 17–21; 71 (October 1987): 19–23.

Prendergast, J. **"Tomorrow's Energy Today."** *Civil Engineering (Am Soc Civ Eng)* 60 (March 1990): 76–79.

Reece, N. S., and G. R. Nuss. **"Renewable Energy: Energy for Today and Tomorrow."** *ASTM Standardization News* 18 (September 1990): 32–37.

Sawyer, Stephen W. **"State Renewable Energy Policy: Program Characteristics, Protection, Needs."** *State and Local Government Review* 17 (Winter 1985): 147–154.

"Social and Private Costs of Alternative Energy Technologies." *Contemporary Policy Issues* 81 (July 1990): 1–307.

U.S. Congress. House. Committee on Banking, Finance, and Urban Affairs. Subcommittee on Economic Stabilization. *Contributions of Renewable and Alternative Fuels to Long-Term U.S. Energy Security: Hearing, February 21, 1990.* 101st Congress, 2nd session. Washington, DC: GPO, 1990, 216 p.

U.S. Congress. House. Committee on Energy and Commerce. Subcommittee on Energy and Power. *Renewable Energy Technologies: Hearing, April 27, 1988, on H.R. 4226, a Bill to Promote the Development and Commercialization of Renewable Energy and Energy Conservation.* 100th Congress, 2nd session. Washington, DC: GPO, 1989, 227 p.

U.S. Congress. House. Committee on Energy and Commerce. Subcommittee on Fossil and Synthetic Fuels. *Alcohol Fuel and Lead Phasedown: A Report, August 1986.* 99th Congress, 2nd session. Washington, DC: GPO, 1986, 108 p.

U.S. Congress. House. Committee on Science, Space and Technology. Subcommittee on Energy Research and Development. *Renewable Energy Development: Hearings, July 8–9, 1987.* 100th Congress, 1st session. Washington, DC: GPO, 1987, 321 p.

————. *Hydrogen: Fuel of the Future: Hearing, September 23, 1987.* 100th Congress, 1st session. Washington, DC: GPO, 1988, 195 p.

————. *H.R. 4226, Renewable Energy and Energy Conservation Commercialization and Development Act: Hearings, June 30, 1988.* 100th Congress, 2nd session. Washington, DC: GPO, 1988, 133 p.

U.S. Congress. House. *Review of the Role of Ethanol in the 1990s; Joint Hearing, May 11, 1988, Before the Subcommittee on Forests, Family Farms, and*

Energy and the Subcommittee on Wheat, Soybeans, and Feed Grains of the Committee on Agriculture and the Subcommittee on Energy and Power of the Committee on Energy and Commerce. 100th Congress, 2nd session. Washington, DC: GPO, 1988, 156 p.

U.S. Congress. Senate. Committee on Energy and Natural Resources. *Renewable and Other Alternative Energy Sources: Hearing, November 1, 1990, Before the Subcommittees on Energy Research and Development and Mineral Resources Development and Production.* 101st Congress, 2nd session. Washington, DC: GPO, 1991, 127 p.

U.S. Congress. Senate. Committee on Energy and Natural Resources. Subcommittee on Energy Research and Development. *Renewable Energy Technologies: Hearings, March 24 and 26, 1987, on the Current Status of the Renewable Energy Technologies in Our National Energy Policy, Fiscal Years 1988–89.* 100th Congress, 1st session. Washington, DC: GPO, 1987, 680 p.

————. *Solar Development Initiative Act of 1987 and the Renewable Energy and Energy Conservation Competitiveness Act of 1987: Hearing, August 6, 1987, on S.1320 [and] S.1554.* 100th Congress, 1st session. Washington, DC: GPO, 1987, 173 p.

————. *Fuel Cell Research and Development and Utilization Policy, and Hydrogen Research and Development: Hearing, September 23, 1987, on S.1294, S.1295 [and] S.1296.* 100th Congress, 1st session. Washington, DC: GPO, 1988, 229 p.

U.S. Congress. Senate. Committee on Environment and Public Works. Subcommittee on Environmental Protection. *Alternative Fuels: Hearings, January 11, 1990.* 101st Congress, 2nd session. Washington, DC: GPO, 1990, 352 p.

U.S. General Accounting Office. *Alcohol Fuels: Impacts from Increased Use of Ethanol Blended Fuels: Report to the Chairman, Subcommittee on Energy and Power, Committee on Energy and Commerce, House of Representatives.* Gaithersburg, MD: 1990, 43 p.

Walls, Margaret A., and others. *Ethanol Fuel and Non-Market Benefits: Is a Subsidy Justified?* Washington, DC: Resources for the Future, Energy and Natural Resources Division, 1989, 34 p.

Williams, Susan, and others. **"Renewing Renewable Energy."** *Issues in Science and Technology* 6 (Spring 1990): 64–70 p.

Hydroelectric Power

Attey, John W., and Drew R. Liebert. **"Clean Water, Dirty Dams: Oxygen Depletion and the Clean Water Act [of 1982]: Legislative and Regulatory Policies That Would Prevent Environmental Degradation Caused by Dams and Hydroelectric Power Plants."** *Ecology Law Quarterly* 11:4 (1984): 703–729.

Blumm, Michael C. **"A Trilogy of Tribes v. FERC: Reforming the Federal Role in Hydropower Licensing."** *Harvard Environmental Law Review* 10:1 (1986): 1–59.

Lambert, Jeremiah D., and James R. O'Neill. **"Privatization of Municipal Hydroelectric Facilities Under Current Law."** *Public Utilities Fortnightly* 121 (February 4, 1988): 11–17.

Reisner, Marc. **"America's Newest Old Energy Source: Hydro Power; Wherever Water Flows and Drops in Appreciable Distance, Someone, Somewhere, Is Contemplating a Dam."** *Amicus Journal* 6 (Spring 1985): 42–52.

Swainson, Neil A. **"The Columbia River Treaty: Where Do We Go from Here?"** *Natural Resources Journal* 26 (Spring 1986): 243–259.

U.S. Congress. Senate. Committee on Energy and Natural Resources. Subcommittee on Water and Power. *Hydroelectric Fairness Act of 1989: Hearing, May 18, 1989, on S. 635 to Prevent the Unintended Licensing of Federally Nonjurisdictional Pre-1935 Unlicensed Hydroelectric Projects.* 101st Congress, 1st session. Washington, DC: GPO, 1989, 158 p.

————. *Hydroelectric Regulation Under Federal Power Act: Hearing, September 28, 1989.* 101st Congress, 1st session. Washington, DC: GPO, 1990, 587 p.

U.S. Federal Energy Regulatory Commission. *Hydroelectric Power Resources of the United States, January 1, 1984: Developed and Undeveloped.* Washington, DC: GPO, 1985, 245 p.

Whittaker, M. Curtis. **"The Federal Power Act and Hydropower Development: Rediscovering State Regulatory Powers and Responsibilities."** *Harvard Environmental Law Review* 10:1 (1986): 135–187.

Young, John E. **"Aluminum's Real Tab."** *World Watch* 5 (March/April 1992): 26–33.

Geothermal

Bradbrook, Adrian J. **"The Contents of New Thermal Legislation."** *Journal of Energy and Natural Resources Law* 5:2 (1987): 81–108.

Cataldi, R. **"World Geothermal Development: The Present Situation and Opportunities for the Future."** *Impact of Science on Society* 37:4 (1987): 337–350.

Pasqualetti, M. J., and Mark Dellinger. **"Hazardous Waste from Geothermal Energy: A Case Study."** *Journal of Energy and Development* 13 (Spring 1988): 275–295.

Rummel, Fritz. **"Geothermal Energy: An Alternative or Additional Potential Energy Source?"** *Applied Geography and Development* 25 (1985): 7–24.

U.S. Congress. Senate. Committee on Energy and Natural Resources. Subcommittee on Energy and Mineral Resources. *Geothermal Energy Development in Nevada's Great Basin: Hearing, April 17, 1984, to Examine the Current Status and Future Needs of Nevada's Geothermal Energy Industry.* 98th Congress, 2nd session. Washington, DC: GPO, 1984, 280 p.

————. *Geothermal Steam Act Amendments of 1987: Hearing, July 14, 1987: on S.1006.* 100th Congress, 1st session. Washington, DC: GPO, 1988, 277 p.

U.S. Energy Information Administration, Office of Coal, Nuclear, Electricity, and Alternative Fuels. *Geothermal Energy in the Western United States and Hawaii: Resources and Projected Electricity Generation Supplies.* Washington, DC: GPO, 1991, 70 p.

Solar Energy

Berg, Mark R., and Mark L. Hassett. **"Conservation and Solar Energy in Residential and Commercial Buildings: The Northern Industrialized USA [Michigan]."** *Energy Policy* 12 (March 1984): 93–101.

Bluestone, Mimi. **"Solar Power: Alive, Well and Almost Making Money."** *Business Week* (July 18, 1988): 132–133.

Bradbrook, Adrian J. **"Australian and American Perspectives on the Protection of Solar and Wind Access."** *Natural Resources Journal* 28 (Spring 1988): 229–267.

————. "The Role of the Courts in Advancing the Use of Solar Energy." *Journal of Energy Law and Policy* 9:2 (1989): 135–175.

Brady, M. E. "Solar Power Subsidization." *Energy* 12 (January 1987): 67–73.

Crouch, Gregory. "The Sun Also Rises—Again: When Oil Prices Skyrocketed, the Solar Industry Soared: But the Switch to the Sun Wasn't for Real; Come 1995, Though, Solar Power Will Be Back—for Good." *Florida Trend* 29 (March 1989): 77–80+.

Flavin, Chris, and Nicholas Lenssen. "Policies for a Solar Economy." *Energy Policy* 20 (March 1992): 245–256.

Merritt, C., and R. Carter. "Sun Sees Its Future—and It's Bright." *Coal* 26 (October 1989): 34–47.

Peck, Louis. "Here Comes the Sun: Environmental Protection and Economic Development Are Working Together on Behalf of a Solar Future." *Amicus Journal* 12 (Spring 1990): 27–32.

Rick, Daniel, and J. David Roessner. "Tax Credits and U.S. Solar Commercialization Policy." *Energy Policy* 18 (March 1990): 186–198.

"Solar Thermal Power." *Natural Resources Journal* 25 (October 1985 supplement): 1097–1157, 4 articles.

Thayer, Robert L., Jr. "Solar Access Control Strategies for Vegetation in Existing and New Developments." *Journal of Architectural and Planning Research* 3 (August 1986): 199–217.

Thorton, J. P., and others. "Photovoltaics—Today's Reality, Tomorrow's Promise." *Energy Engineering* 87:3 (1990): 63–79.

U.S. Congress. Senate. Committee on Energy and Natural Resources. Subcommittee on Energy Research and Development. *Conservation and Solar Energy Research and Development: Hearing, July 13, 1987, on the Current Status of Solar Energy Research and Development in the Federal Government.* 100th Congress, 1st session. Washington, DC: GPO, 1987, 329 p.

Weinberg, C. J., and R. H. Williams. "Energy from the Sun." *Scientific American* 263 (September 1990): 146–155.

Biomass

Carioca, J. O. B., and others. **"Energy from Biomass."** *Impact of Science on Society* 37:4 (1987): 305–315.

Goltfried, Robert R. **"Can Energy Cane Stern the Tide?"** *Social and Economic Studies* 36 (September 1987): 177–202.

Gowen, Marcia M. **"Biofuel v. Fossil Fuel Economics in Developing Countries."** *Energy Policy* 17 (October 1989): 455–470.

Heid, Walter G., Jr. *Turning Great Plains Crops Residues and Other Products into Energy.* Agriculture Economic Report No. 523. Washington, DC: U.S. Department of Agriculture, Economic Research Service, 1984, 37 p.

Herendeen R., and S. Brown. **"A Comparative Analysis of Net Energy from Woody Biomass."** *Energy* 12 (January 1987): 75–84.

Jones, Harold B., Jr., and E. A. Ogden. *Energy Potential from Livestock and Poultry Waste in the South.* Agriculture Economic Report No. 522. Washington, DC: U.S. Department of Agriculture, Economic Research Service, 1984, 40 p.

Lichtman, Rob. **"Toward the Diffusion of Rural Energy Technologies; Some Lessons from the Indian Biogas Program."** *World Development* 15 (March 1987): 342–374.

Makansi, J. **"Co-combustion: Burning Biomass, Fossil Fuels Together Simplifies Waste Disposal, Cuts Fuel Cost."** *Power* 131 (July 1987): 11–18.

New Jersey General Assembly. County Government and Regional Authorities Committee. *Committee Meeting on Assembly Bill 1778 (Provides for a Resource Recovery Investment Tax on Solid Waste Disposal at Sanitary Landfills): Held: Trenton, New Jersey, May 14, 1984.* Trenton, NJ: 1984, 5 p.

Reddy, A. K. N., and J. Goldemberg. **"Energy for the Developing World."** *Scientific American* 263 (September 1990): 110–118.

"Solid Fuels/Biomass; Products and Services." *Independent Energy* 19 (December 1989): 93–100.

Strauss, C. H., and L. L. Wright. **"Woody Biomass Production Costs in the United States: An Economic Survey of Commercial Populous Plantation Systems."** *Solar Energy* 45:2 (1990): 105–110.

U.S. Tennessee Valley Authority. National Fertilizer Development Center. *Biomass Fuels Update II*. Bulletin Y-184. Muscle Shoals, AL: 1984, 58 p.

Winter, R. M., and others. **"Biomass Combustion: Relationship Between Pollutants Formation and Fuel Composition."** *Tappi Journal* 72 (April 1989): 139–145.

Zerbe, John I. **"Biofuels: Production and Potential."** *Forum for Applied Research and Public Policy* 3 (Winter 1988): 38–44.

Wind

Bradbrook, Adrian J. **"Australian and American Perspectives on the Protection of Solar and Wind Access."** *Natural Resources Journal* 28 (Spring 1988): 229–267.

Eide, David. **"Bright Prospects for Wind Energy."** *Strategic Planning Energy and Environment* 11 (Summer 1991): 40–52.

Gray, Peter, and Erica Rosenberg. **"Turning On Wind Power: Enlightened Public Policy Could Tap This Cheap and Inexhaustible Resource, Saving Money and the Environment."** *Environmental Forum* 8 (September/October 1991): 16–22.

Jarvis, Michaela. **"Power from Thin Air."** *California Journal* 20 (July 1989): 278–283.

Kellett, Jon. **"The Environmental Impact of Wind Energy Developments."** *Town Planning Review* 61 (April 1990): 139–155.

Pritchard, C., and R. Low. **"A Self-Regulating Heat Pump to Utilize Wind and Wave Energy Sources."** *Energy Sources* 12:1 (1990): 15–24.

Shea, Cynthia Pollock. **"Harvesting the Wind: Launched in California, Wind Power Is Now Spanning the Globe."** *World Watch* 1 (March/April 1988): 12–17.

Starrs, Thomas A. **"Legislative Incentives and Energy Technologies: Government's Role in the Development of the California Wind Energy Industry."** *Ecology Law Quarterly* 115:1 (1988): 103–158.

Ocean Power

Baker, A. C. **"Tidal Power."** *IEE Proceedings Part A* 134 (May 1987): 392–398.

Kindl, John Warren. **"Ocean Thermal Energy Conversion."** *Georgia Journal of International and Comparative Law* 14 (Winter 1984): 1–27.

Lander, James F., and Patricia A. Lockridge. *United States Tsunamis, 1690–1988.* Washington, DC: U.S. National Oceanic and Atmospheric Administration, 1989, 265 p.

Lennard, D. E. **"Ocean Thermal Energy Conversion—Past Progress and Future Prospects."** *IEE Proceedings Part A* 134 (May 1987): 381–391.

Penney, T. R., and D. Bharathan. **"Power from the Sea."** *Scientific American* 256 (January 1987): 86–92.

Watts, S. **"Marine Directorate to Investigate Energy from Oceans."** *New Scientist* 120 (December 7, 1988): 24–31.

Wood as Fuel

Dewees, Peter A. **"The Woodfuel Crisis Reconsidered: Observations on the Dynamics of Abundance and Scarcity."** *World Development* 17 (August 1989): 1159–1172.

Lipfert, F. W., and J. Lee. **"Air Pollution Implications of Increasing Residential Firewood Use."** *Energy* 10 (January 1985): 17–33; Discussion 13 (December 1988): 883 p.

Peacock, R. D. **"Wood Heating Safety Research: An Update."** *Fire Technology* 23 (November 1987): 292–312.

Perlack, R. D., and W. A. Geyer. **"Wood Energy Plantation Economics in the Great Plains."** *Journal of Energy Engineering* 113 (December 1987): 92–101.

Tonn, B. E., and D. L. White. **"Residential Demand for Wood: Results from the U.S. Pacific Northwest."** *Energy* 12 (December 1987): 1265–1274.

————. **"Patterns of Residential Wood and Electricity Use: Results from the Hood River Conservation Project."** *Energy* 13 (June 1988): 485–497.

————. **"Residential Wood Consumption for Space Heating in the Pacific Northwest: Difficulties for Electric Utilities?"** *Energy Policy* 18 (April 1990): 283–292.

Traynor, G. W., and others. **"Indoor Air Pollution Due to Emissions from Wood-Burning Stoves."** *Environmental Science and Technology* 21 (July 1987): 691–697.

Nuclear Power

General

Benedict, Manson. **"Nuclear Power in the World Today."** *Proceedings of the American Philosophical Society* 129 (December 1985): 456–475.

Badansky, David. **"Global Warming and Oil: Can Nuclear Power Make a Difference?"** *Electricity Journal* 4 (January/February 1991): 34–44.

Campbell, John L. **"The State, Capital Formation, and Industrial Planning: Financing Nuclear Energy in the United States and France."** *Social Science Quarterly* 67 (December 1986): 707–721.

Clarke, Lee. **"The Origins of Nuclear Power: A Case of Institutional Conflict."** *Social Problems (Social Study Social Problems)* 32 (June 1985): 474–487.

Collingridge, D. **"Lessons of Nuclear Power: U.S. and U.K. History."** *Energy Policy* 12 (March 1984): 46–67.

Häfele, Wolf. **"Energy from Nuclear Power."** *Scientific American* 263 (September 1990): 136–144.

Hodel, Donald. **"The Place of Nuclear Power in the Reagan Administration's Energy Policy."** *Public Utilities Fortnightly* 114 (October 11, 1984): 21–25.

Jones, L. R. **"The WPPSS Default: Trouble in the Municipal Bond Market [Causes and Consequences of the Default on Washington Public Power Supply System Nuclear Plants 4 and 5]."** *Public Budgeting and Finance* 4 (Winter 1984): 60–77.

Kriz, Margaret E. **"Boasting Nuclear: The Beleagured Nuclear Power Industry Has the Bush Administration On Its Side As It Tries to Change the Rules to Let Electric Utilities Get Back in the Business of Building Nuclear Power Plants."** *National Journal* 23 (February 23, 1991): 440–445.

Lanouette, William. **"Atomic Energy, 1945–1985."** *Wilson Quarterly* 9 (Winter 1985): 90–131.

Lester, Richard K. **"Organization, Structure, and Performance in the U.S. Nuclear Power Industry."** *Energy Systems and Policy* 9:4 (1986): 335–384.

Lowen, Rebecca S. **"Entering the Atomic Power Race: Science, Industry, and Government."** *Political Science Quarterly* 102 (Fall 1987): 459–479.

Mariotte, Michael. **"The Resurgence of Nuclear Power in America."** *Multinational Monitor* 10 (January/February 1989): 9–13.

Murray, Jan. **"Can Nuclear Energy Contribute to Slowing Global Warming?"** *Energy Policy* 18 (July/August 1990): 494–499.

"Nuclear Power and Land Use." *Land Use Policy* 5 (January 1988): 2–74.

Pasztor, Janos. **"What Role Can Nuclear Power Play in Mitigating Global Warming?"** *Energy Policy* 19 (March 1991): 98–109.

Szöke, Abraham, and Ralph W. Moir. **"A Practical Route to Fusion Power: Small Underground Nuclear Explosions Could Supply the World's Electricity for Centuries to Come: Unlike Other Forms of Fusion, This Technology Is Feasible and Affordable Now."** *Technology Review* 94 (July 1991): 20–27.

U.S. Congress. House. Committee on Energy and Commerce. Subcommittee on Energy and Power. *Renewable Energy Technologies Hearing, April 26, 1989, on H.R. 1216, a Bill to Provide Federal Assistance and Leadership to a Program of Research, Development and Demonstration of Renewable Energy and Energy Efficiency Technologies, and for Other Purposes.* 101st Congress, 1st session. Washington, DC: GPO, 1989, 153 p.

U.S. Congress. Office of Technology Assessment. *Starpower: The U.S. and International Quest for Fusion Energy.* Washington, DC: GPO, 1987, 237 p.

U.S. Congress. Senate. Committee on Environment and Public Works. Subcommittee on Nuclear Regulation. *The Role of Nuclear Energy in Meeting Future Electricity Demands: Hearing, August 1, 1990.* 101st Congress, 2nd session. Washington, DC: GPO, 1990, 101 p.

U.S. Energy Information Administration. Office of Coal, Nuclear, Electricity, and Alternate Fuels. *Commercial Nuclear Power, 1986: Prospects for the United States and the World.* Washington, DC: GPO, 1986, 142 p.

————. *Commercial Nuclear Power, 1988: Prospects for the United States and the World.* Washington, DC: GPO, 1988, 130 p.

Williams, Susan, and others. **"Renewing Renewable Energy."** *Issues in Science and Technology* 6 (Spring 1990): 64–70.

"World Status Report: Nuclear Power." *Energy Economist* (January 1990): 11–18.

Zillman, Donald. **"Death at Early Middle Age: Reflections on the Future of Nuclear Energy in the United States."** *Journal of Energy and Natural Resources* 7:1 (1989): 34–41.

Public Opinion

DeBoer, Connie, and Ineke Catsburg. **"The Impact of Nuclear Accidents on Attitudes Toward Nuclear Energy."** *Public Opinion Quarterly* 52 (Summer 1988): 254–261.

Flavin, Christopher. **"The Case Against Reviving Nuclear Power Is Now Being Pushed as a Cure for Global Warming: Here Is Why That Thinking Is Wrong."** *World Watch* 1 (July/August 1988): 27–35.

Gamson, William A., and Andre Modigliani. **"Media Discourse and Public Opinion on Nuclear Power: A Constructionist Approach."** *American Journal of Sociology* 95 (July 1989): 1–37.

"Nuclear Decay: Bombs Away—Nemesis for Nuclear Energy." *South* (September/October 1991): 9–20.

Taylor, John. **"Nuclear Power in the United States: A Current Assessment."** *Electricity Journal* 4 (January/February 1991): 26–33.

Technology

Campbell, John L. **"The State and the Nuclear Waste Crisis: An Institutional Analysis of Policy Constraints."** *Social Problems (Social Study Social Problems)* 34 (February 1987): 18–33.

Crane, Alan T., and Mark Mills. **"Nuclear Power's Future: Technology and Policy."** *Forum for Applied Research and Public Policy* 1 (Fall 1986): 8–19.

Gaunt, John, and Neil J. Numark. *Decommissioning of Nuclear Power Facilities.* Washington, DC: International Bank for Reconstruction and Development, Industry and Energy Department, 1990, 37 p.

U.S. Congress. House. Committee on Energy and Commerce. Subcommittee on Energy Conservation and Power. *Development of Nuclear Power*

Fuel Cycles: Report November 1984. 98th Congress, 2nd session. Washington, DC: GPO, 1984, 136 p.

————. *Price-Anderson Provisions of the Atomic Energy Act of 1954 to Establish Liability and Indemnification for Nuclear Incidents Arising Out of Federal Storage, Disposal, or Related Transportation of High-Level Radioactive Waste and Spent Nuclear Fuel.* 99th Congress, 2nd session. Washington, DC: GPO, 1987, 344 p.

U.S. Congress. Senate. Committee on Energy and Natural Resources. Subcommittee on Energy Research and Development. *Advanced Reactor Development Program: Hearing, May 24, 1988.* 100th Congress, 2nd session. Washington, DC: GPO, 1988, 266 p.

Accidents

Ewart, Elizabeth M., and Stanley B. Jacobs. **"The Future of Nuclear Power in the United States: Fission and Fusion."** *Public Utilities Fortnightly* 119 (April 30, 1987): 28–33.

Glazer, Sarah. **"After Chernobyl: Nuclear Power Safety in Doubt."** *Editorial Research Reports* (December 5, 1986): 895–912.

Ramberg, Bennett. **"Learning from Chernobyl."** *Foreign Affairs* 65 (Winter 1986/1987): 304–328.

Tomain, Joseph, and Constance Dowd Burton. **"Nuclear Transition: From Three Mile Island to Chernobyl."** *William and Mary Law Review* 28 (Spring 1987): 363–437.

U.S. Congress. Senate. Committee on Energy and Natural Resources. *The Chernobyl Accident: Hearing, June 19, 1986, on the Chernobyl Accident and Implications for the Domestic Nuclear Industry.* 99th Congress, 2nd session. Washington, DC: GPO, 1986, 343 p.

Laws and Regulations

Ahrari, M. E. **"Congress, Public Opinion, and Synfuels Policy."** *Political Science Quarterly* 102 (Winter 1987/1988): 589–606.

Goodman, Marshall R., and Fredric Andes. **"The Politics of Regulatory Reform and Future Direction of Nuclear Energy Policy."** *Policy Studies Review* 5 (August 1985): 111–121.

Kaufman, Allen. **"Public Policy and Synthetic Fuels; Challenges for Business Solidarity,"** in Preston, Lee, ed. *Research in Corporate Social Performance and Policy, 1984.* Greenwich, CT: JAI Press, 1984, 187–211.

Siegel, John R., and John O. Sillin. **"Revitalizing Nuclear Power: The Case for Deregulation."** *Public Utilities Fortnightly* 117 (January 23, 1986): 18–25.

Straubel, Michael S. **"Space Borne Nuclear Power Sources: The Status of Their Regulation."** *Valparaiso University Law Review* 20 (Winter 1986): 187–218.

U.S. Congress. House. Committee on Energy and Commerce. *Methanol as Transportation Fuel: Hearings, April 4 and 25, 1984, Before the Subcommittee on Fossil and Synthetic Fuels and the Subcommittee on Energy Conservation and Power, on H.R. 4855 and H.R. 5075, Bills to Develop a National Methanol Energy Policy, to Amend the Energy Policy and Conservation Act, to Improve the Environment Through the Use of Methanol, and for Other Purposes.* 98th Congress, 2nd session. Washington, DC: GPO, 1984, 372 p.

U.S. Congress. House. Committee on Energy and Commerce. Subcommittee on Energy Conservation and Power. *Nuclear Regulatory Commission Sunshine Act Regulations: Hearings, May 21, 1985.* 99th Congress, 1st session. Washington, DC: GPO, 1985, 126 p.

U.S. Congress. House. Committee on Energy and Commerce. Subcommittee on Fossil and Synthetic Fuels. *Synthetic Fuels Policy: Hearings, October 4, 1983–June 18, 1984.* 98th Congress, 1st and 2nd sessions. Washington, DC: GPO, 1984, 469 p.

————. *Methanol Fuel: Kicking the Gasoline Habit: Proceedings of a Seminar May 2, 1984.* 98th Congress, 2nd session. Washington, DC: GPO, 1984, 88 p.

U.S. Congress. House. Committee on Foreign Affairs. *United States-Japan Nuclear Cooperation Agreement: Hearings, December 16, 1987 and March 2, 1988.* 100th Congress, 1st and 2nd sessions. Washington, DC: GPO, 1988, 706 p.

U.S. Congress. House. Committee on Science and Technology. Subcommittee on Energy Development and Applications. *The Status of Synthetic Fuels and Cost-Shared Energy R & D Facilities: Hearings, June 6–13, 1984.* 98th Congress, 2nd session. Washington, DC: GPO, 1984, 405 p.

U.S. Congress. House. Committee on the Judiciary. Subcommittee on Administrative Law and Governmental Relations. *State and Local Law-Enforcement Compensation Act: Hearing, June 2, 1988, on H.R. 3711.* 100th Congress, 2nd session. Washington, DC: GPO, 1988, 115 p.

U.S. Congress. House. Committee on Ways and Means. Subcommittee on Oversight. *Tax Incentives for the Production and Use of Ethanol Fuels:*

Hearing, July 9, 1984. 98th Congress, 2nd session. Washington, DC: GPO, 1984, 283 p.

U.S. Congress. Senate. Committee on Energy and Natural Resources. *Nuclear Facility Standardization Act of 1986: Hearing, April 22, 1986, on S. 2073, a Bill to Encourage the Standardization of Nuclear Powerplants, to Improve the Nuclear Licensing and Regulatory Process, to Amend the Atomic Energy Act of 1954 and for Other Purposes.* 99th Congress, 2nd session. Washington, DC: GPO, 1986, 483 p.

————. *Potential Alternative Transportation Fuels Other Than Methanol: Hearings, November 14, 1989.* 101st Congress, 1st session. Washington, DC: GPO, 1990, 304 p.

U.S. Congress. Senate. Committee on Environment and Public Works. Subcommittee on Nuclear Regulation. *Price-Anderson Act Amendment of 1987.* 100th Congress, 1st session. Washington, DC: GPO, 1987, 390 p.

————. *Nuclear Regulation Reorganization and Reform Act of 1989: Hearing, July 19, 1989, on S. 946, a Bill to Reorganize the Functions of the Nuclear Regulatory Commission to Promote More Effective Regulation of Atomic Energy for Peaceful Purposes.* 101st Congress, 1st session. Washington, DC: GPO, 1989, 107 p.

U.S. Laws and Statutes. *Compilation of Selected Energy-Related Legislation: Vol 1, Nuclear Energy.* 101st Congress, 1st session. Washington, DC: GPO, 1989, 421 p.

Victor, Kirk. **"The Nuclear Turn-on: The Bush Administration in Crafting a New National Energy Strategy, Seems Determined to Try to Sell the American People On the Need for More Nuclear Power Plants."** *National Journal* 21 (September 9, 1989): 2196–2200.

Fusion

McKinley, K. R., and others. **"Hydrogen Fuel from Renewable Resources."** *Energy Sources* 12:2 (1990): 105–110.

U.S. Congress. House. Committee on Science, Space and Technology. Subcommittee on Investigations and Oversight. *Fusion Energy Program: Status and Direction: Hearing, October 3–26, 1989.* 101st Congress, 1st session. Washington, DC: GPO, 1990, 809 p.

U.S. Congress. Senate. Committee on Energy and Natural Resources. Subcommittee on Energy Research and Development. *DOE's Magnetic Fusion Program: Hearing, June 14, 1989, on the Department of Energy's Magnetic Fusion Research and Development and Demonstration.* 101st Congress, 1st session. Washington, DC: GPO, 1989, 241 p.

Selected Journal Titles

The following journals publish articles on many aspects of energy. Because of the many types of energy and the rapidly changing technical and economic conditions, energy information is found in a wide variety of journals. New journals are continually appearing. For more journals and additional information please consult:

Ulrich's International Periodicals Directory, 1991–1992. 30th edition. New York: R. R. Bowker, 1991, 3 Vols.

Information on the journals listed is arranged in the following manner:

Journal Title

1. Editor
2. Year first published
3. Frequency of publication
4. Code
5. Special features
6. Address of publisher

Amicus Journal

1. Francesca Lyman
2. 1979
3. Quarterly
4. ISSN 0276-7201
5. Bk. rev., illus., index, cum. index
6. Natural Resources Defense Council Inc.
   40 W. 20th Street
   New York, NY 10011

Editorial Research Reports

1. Marcus Rosenbaum
2. 1923

3. 4 issues per month
4. ISSN 0013-0958
5. Bk. rev., charts, index
6. Congressional Quarterly, Inc.
   1414 22nd Street, NW
   Washington, DC 20037

Electricity Journal

1. Robert O. Marritz
2. 1988
3. 10 issues per year
4. ISSN 1040-6190
5. Bk. rev.
6. Robert O. Marritz
   2121 Fourth Avenue, Suite 635
   Seattle, WA 98121

Energy

1. S. S. Penner
2. 1976
3. Monthly
4. ISSN 0360-5442
5. Adv., bk. rev., abstr., illus., stat., index
6. Pergamon Press, Inc.
   Journals Division
   Maxwell House
   Fairview Park
   Elmsford, NY 10523

Energy Economist

1.
2. 1981
3. Monthly
4. ISSN 0262-7108
5.
6. Financial Times Business Information, Ltd.
   Tower House
   Southampton Street
   London WC2E 7HA
   England

Energy Engineering

1. Anna Williams
2. 1904
3. Bimonthly
4. ISSN 0199-8595
5. Adv., bk. rev., charts, illus., tr. lit. index
6. Association of Energy Engineers
   Fairmont Press, Inc.
   700 Indian Trail
   Lilburn, GA 30247

Energy Policy

1. Nicky France
2. 1973
3. 10 issues per year
4. ISSN 0301-4215
5. Adv., bk. rev., charts, illus., stat., index
6. Butterworth-Heinemann, Ltd.
   P.O. Box 63
   Westbury House, Bury Street
   Guilford, Surrey GU2 5BH
   England

Energy Systems and Policy

1. S. William Gouse, Jr.
2. 1973
3. Quarterly
4. ISSN 090-8347
5. Adv., abstr., index
6. Taylor & Francis
   1900 Frost Road, Suite 101
   Bristol, PA 19007-1598

Forum for Applied Research and Public Policy

1. Daniel Schaffer
2. 1986
3. Quarterly
4. ISSN 0887-8218
5. Adv., bk. rev.

6. University of North Carolina Press
   Forum, Box 2288
   Chapel Hill, NC 27515-9979

Issues in Science and Technology

1. Steven J. Marcus
2. 1984
3. Quarterly
4. ISSN 0748-5492
5. Adv., bk. rev., index
6. National Academy of Sciences
   2101 Constitution Avenue, NW
   Washington, DC 20418

Journal of Energy and Development

1. Dorothea H. El Mallakh
2. 1975
3. Twice annually
4. ISSN 0361-4476
5. Adv., bk. rev., charts, stat.
6. International Research Center for Energy and Economic
   Development
   206 Business Building
   University of Colorado
   Campus Box 418
   Boulder, CO 80309-0418

Journal of Energy, Natural Resources and Environmental Law

1. Editorial Board
2. 1980
3. Twice annually
4.
5. Bk. rev., index
6. University of Utah
   College of Law
   Salt Lake City, UT 84112
   Formerly (until 1990): **Journal of Energy Law and Policy**
   (ISSN 0275-9926)

Journal of Energy Resources Technology

1. J. S. Chung
2. 1979

3. Quarterly
4. ISSN 0195-0738
5.
6. American Society of Mechanical Engineers
   345 East 47th Street
   New York, NY 10017

Journal of Policy Analysis and Management

1. David L. Weimer
2. 1981
3. 4 issues per year
4. ISSN 0276-8739
5. Adv., index
6. Association for Public Policy Analysis and Management
   John Wiley & Sons, Inc.
   605 Third Avenue
   New York, NY 10158-0012

National Journal

1. Richard S. Frank
2. 1969
3. Weekly
4. ISSN 0360-4217
5. Adv., bk. rev., charts, illus., index
6. National Journal, Inc.
   1730 M Street, NW, Suite 1100
   Washington, DC 20036

Natural Resources Journal

1. Albert E. Utton
2. 1961
3. Quarterly
4. ISSN 0028-0739
5. Adv., bk. rev., charts, index, cum. index every 10 years
6. University of New Mexico
   School of Law
   1117 Stanford NE
   Albuquerque, NM 87131

OPEC Bulletin

1. Keith Jinks
2. 1969

3. 10 issues per year
4. ISSN 0474-6279
5. Bk. rev., charts, illus., stats., index
6. Organization of the Petroleum Exporting Countries
   Publications
   Public Information Department
   Obere Donaustrasse 93
   A-1030 Vienna
   Austria

Petroleum Economist

1. Ian Bourne
2. 1934
3. Monthly
4. ISSN 0306-395X
5. Adv., bk. rev., charts, mkt., stat., index
6. Petroleum Economist, Ltd.
   P.O. Box 105
   25-31 Ironmonger Row
   London ECIV 3PN
   England

Power (New York)

1. Robert G. Schwieger
2. 1882
3. Monthly
4. ISSN 0032-5929
5. Adv., bk. rev., abstr., charts, illus., stat., tr. lit., index
6. McGraw-Hill, Inc.
   1221 Avenue of the Americas
   New York, NY 10020

Public Utilities Fortnightly

1. Cheryl Romo
2. 1929
3. 24 issues per year
4. ISSN 0033-3808
5. Adv., bk. rev., index semi-annually
6. Public Utilities Reports, Inc.
   2111 Wilson Boulevard, Suite 200
   Arlington, VA 22201

Resources

1. Nancy Money
2. 1982
3. Quarterly
4. ISSN 0714-5918
5.
6. Canadian Institute of Resources Law
   Room 430, Bio-Sciences Building
   Faculty of Law
   University of Calgary
   Calgary, Alberta T2N IN4
   Canada

Resources Policy

1. Roderick G. Eggert
2. 1975
3. Quarterly
4. ISSN 0301-4207
5. Bk. rev., abstr., illus.
6. Butterworth-Heinemann, Ltd.
   P.O. Box 63
   Westbury House, Bury Street
   Guilford, Surrey GU2 5BH
   England

Scientific American

1. Jonathan Piel
2. 1845
3. Monthly
4. ISSN 0036-8733
5. Adv., bk. rev., illus., index, cum. index
6. Scientific American, Inc.
   415 Madison Avenue
   New York, NY 10017

Solar Energy

1. John Duffie
2. 1957
3. 12 issues per year
4. ISSN 0038-092X

5. Adv., bk. rev., abstr., bibl., charts, illus.
6. Pergamon Press, Inc.
   Journals Division
   Maxwell House
   Fairview Park
   Elmsford, NY 10523

Solar Today

1. Maureen McIntyre
2. 1987
3. Bimonthly
4.
5. Adv., bk. rev.
6. American Solar Energy Society
   2400 Central Avenue, B-1
   Boulder, CO 80301

Strategic Planning for Energy and the Environment

1. F. William Payne
2. 1981
3. Quarterly
4. ISSN 1048-5236
5.
6. Fairmont Press, Inc.
   700 Indian Trail
   Lilburn, GA 30247

Technology Review

1. Jonathan Schlefer
2. 1899
3. 8 issues per year
4. ISSN 0040-1692
5. Adv., bk. rev., illus.
6. Massachusetts Institute of Technology
   Association of Alumni and Alumnae
   W59-200
   Cambridge, MA 02139

6

Films and Videocassettes

THE SELECTED FILMS AND VIDEOCASSETTES PRESENT a wide spectrum of information on energy sources. The selection begins with a number of films and videocassettes about energy and its conservation. These are followed by such specific topics as petroleum, coal, natural gas, hydroelectric, nuclear, renewable, and solar energy. These graphic presentations provide information more vividly than the written word. Thus a film may aid in providing information needed to solve a complex energy problem.

The following sources offer films and videocassettes in English.

AFVA Evaluations, 1991. Fort Atkinson, WI: Highsmith Press, 1991, 340 p.

Educational Film/Video Locator of the Consortium of College and University Media Centers and R. R. Bowker. 4th ed. New York: R. R. Bowker, 1990–1991, 2 Vols.

Film & Video Finder. 3rd ed. Medford, NJ: Plexus Publishing, 1991, 3 Vols.

Films and Video for Mathematics and Physical Sciences. 9th ed. University Park, PA: Audio Visual Sciences, Pennsylvania State University, 1991, 114 p.

Video Rating Guide for Libraries. Santa Barbara, CA: ABC-CLIO, 1990–.

The Video Source Book. 13th ed. Detroit, MI: Research, Inc., 1992, 2 Vols., and supplement.

The following data are provided for each film:

Title of film

Distributor

Phone number, if available

Data on film

Description

Energy

Energy: A Light at the End of the Tunnel
Coronet/MTI Film and Video
Supplementary Education Group
Simon and Schuster Communications
108 Wilmot Road
Deerfield, IL 60015
Phone: (800) 323-5343
Color, 21 minutes, sound, 16mm, 1981.

Our energy dilemma can be solved by developing new energy sources and technologies and making better use of existing fuel supplies. The film examines energy efficient policies and practices applied to transportation, industry, and housing. It shows that conservation is a cost-efficient investment and a cheap source of energy.

Energy: An International Crisis
AIMS Media
6901 Woodley Avenue
Van Nuys, CA 91406
Phone: (818) 785-4111
        (800) 367-2467
Color, 13 minutes, sound, 16mm, ¹/₂" VHS, 1980.

This film discusses the history of energy use. It examines priorities to be used in emergency situations in air, traffic, military, industrial, and agricultural use. It stresses conservation.

Energy: Innovative Alternatives
AIMS Media
6901 Woodley Avenue
Van Nuys, CA 91406
Phone: (800) 367-2467
       (818) 785-4111
Color, 17 minutes, sound, 16mm, 1/2" VHS, 1980.

This film examines new types of motors for alternative fuels, wind power, research on solar energy, bio-process for preparing wood pulp for paper manufacture, and promising new technologies.

Energy: The Dilemma (Second Edition)
Churchill Films
12210 Nebraska Avenue
Los Angeles, CA 90028
Phone: (213) 207-6600
       (200) 334-7830
Fax: (213) 207-1330
Color, 22 minutes, sound, 16mm, 1/2" VHS, 1980.

Examines oil, gas, coal, and nuclear power with respect to potential supplies, rising costs, environmental hazards, and the unique problems of nuclear reactors. Surveys the problems of increasing consumption of fuel and the economic impact of growing oil imports.

Energy Alternatives in Perspective
Iowa State University
121 Pearson Hall
Media Resources Centre
Ames, IA 50011
Color, 78 minutes, sound, 1/2" VHS, 1984.

The film reviews nuclear, coal, oil, solar, and geothermal energy sources discussed during a panel discussion at Iowa State University.

Energy at Work
Churchill Films
12210 Nebraska Avenue
Los Angeles, CA 90025
Phone: (213) 207-6600
       (800) 334-7830
Fax: (213) 207-1330
Color, 13 minutes, sound, 16mm, 1983.

This film shows different kinds of energy, such as heat, mechanical, light, chemical, and nuclear. It explains input and output of energy and how it is transferred from one form to another. It includes animation and experiments by young people.

The Energy Challenge
University of Nebraska
Bureau of Audio-Visual Instruction
421 Nebraska Hall
Lincoln, NE 68508
Color, 30 minutes, sound, 3/4" U-matic, 1980 (each program).

Program 1 shows how the use of petroleum products has increased at the Living History Farms in Iowa. It also shows how the 1970s energy crisis was brought on by shortages of fossil fuels due to embargoes.

Program 2 shows that a great deal of energy is used in personal transportation and in homes and how people are responding to the energy shortage.

Program 3 takes an in-depth look at energy consumed in industry, including agriculture. It shows how businesses have met the energy challenge by innovations in their operations. Also shown is how schools can and do make progress in energy management.

Program 4 shows how we may have to change our energy use habits because of the costs of autos and gasoline. It hopes to make the youth of today become energy savers.

Energy Does Work
Barr Films
12801 Schabarum Avenue
P.O. Box 7878
Irwindale, CA 91706
Phone: (818) 338-7878
Color, 14 minutes, sound, 16mm, 1981.

Illustrates how chemical, thermal, nuclear, electrical, mechanical, and radiant energy has the ability to do work excluding its connection with fossil fuels.

Energy Efficient Homes
ABC Distribution Company
825 Seventh Avenue
New York, NY 10019-6001
Phone: (212) 887-1725
         (212) 887-1731
Color, 30 minutes, sound, Beta, 1/2" VHS, 3/4" U-matic, 1988.

An ABC news report on the costs and benefits of energy efficient housing.

Energy for Tomorrow
EME
Old Mill Plain Road
P.O. Box 2805
Danbury, CT 06813-2805
Phone: (800) 848-2050
Color, 40 minutes, sound, 1/2" VHS, 1991.

Examines existing energy sources as well as potential sources for the future and discusses the costs, benefits, and hazards of each.

The Energy Savers
Disney Educational Productions
500 S. Buena Vista Street
Burbank, CA 91521
Phone: (800) 621-2131
Color, 9 minutes, sound, Beta, 1/2" VHS, EJ, 3/4" U-matic, 1990.

A cartoon showing how energy should be conserved.

Energy To Go Round
Media Guild
11722 Sorrento Valley Road, Suite E
San Diego, CA 92121
Phone: (619) 755-9191
Color, 24 minutes, sound, 1/2" VHS, 1981.

Examines the ways to store energy in the flywheel of a toy truck and in an old pumping engine. Demonstrates research on a modern flywheel system and shows how the earth is a gigantic flywheel in rotation.

Fueling the Future
Video Project
5332 College Avenue, Suite 101
Ashland, CA 94618
Phone: (510) 655-9050
Color, 58 minutes, sound, Beta, 1/2" VHS, 3/4" U-matic, 1988.

A view of the relationship between the American life-style and energy usage, focusing on possible alternatives.

The Time Traveler's Guide to Energy
Disney Educational Productions
500 S. Buena Vista Street
Burbank, CA 91521
Phone: (800) 621-2131
Color, 27 minutes, sound, Beta, 1/2" VHS, EJ, 3/4" U-matic, 1990.

A boy from the past meets a boy from the present, and together they re-create our energy past, present, and the future.

Conservation

Energy: How We Got Where We Are
Moody Institute of Science
12000 E. Washington Boulevard
Whittier, CA 90608
Color, 20 minutes, sound, 16mm, 1981.

This film shows how we must take inventory of our supply of natural resources (oil, gas, and coal) because we are depleting them through our affluent life-style.

Energy: Less is More (Second Edition)
Churchill Films
12210 Nebraska Avenue
Los Angeles, CA 90025
Phone: (213) 207-6600
          (800) 334-7830
Fax: (213) 207-1330
Color, 21 minutes, sound, 16mm, 1980.

Because fossil fuels are diminishing, emphasis is placed on the conservation of energy and the time required to develop renewable sources. Focus is on the areas of transportation, industry, and buildings.

Energy: Saving on the Farm
University of Wisconsin
Bureau of Audio Visual Instruction
P.O. Box 2093
Madison, WI 53701
Color, 26 minutes, sound, 1/2" VHS, 1981.

The film discusses the many uses of energy on farms and how it might be used more efficiently.

Energy Crunch: The Best Way Out
Carousel Film and Video
260 Fifth Avenue
New York, NY 10001
Phone: (212) 663-1660
Color, 51 minutes, sound, 16mm, 1980.

Explores ways in which houses, industries, and automobiles can be made more energy efficient, what local and federal agencies can do to increase energy conservation, and the potential of solar and other alternative energy sources. Shows how Portland, Oregon, has improved mass transit, changed zoning laws to reduce urban sprawl, and implemented weatherization of buildings.

Energy in Physics
Disney Educational Productions
500 S. Buena Vista Street
Burbank, CA 91521
Phone: (800) 621-2131
Color, 20 minutes, sound, Beta, 1/2" VHS, EJ, 3/4" U-Matic, 1990.

This video focuses on the Law of Conservation of Energy, along with the concepts of work equals four times distance, friction, energy efficiency, kinetic and potential energy, and the conversion of energy.

Energy Sources
International Film Bureau
332 S. Michigan Avenue
Chicago, IL 60604
Phone: (312) 427-4545
Color, 19 minutes, sound, 16mm, 1/2" VHS, 1982.

Shows how we have become energy dependent and investigates sources that may supply energy in the future.

The Great Search: Man's Need for Power and Energy
Disney Educational Productions
500 S. Buena Vista Street
Burbank, CA 91521
Phone: (800) 621-2131
Color, 13 minutes, sound, Beta, 1/2" VHS, EJ, 1990.

Details man's search to find efficient sources of energy without upsetting the ecological balance.

Of Energy, Minerals and Man Series
Journal Films
930 Pitner Avenue
Evanston, IL 60202
Phone: (800) 323-5448
       (708) 328-6700
Fax: (708) 328-6706
Color, 27 minutes, sound, Beta, 1/2" VHS, 3/4" U-matic, 1985.

Demonstrates man's use of minerals and examines the role conservation plays with these rapidly depleting resources.

Petroleum

Oil: Episode 1—God Bless Standard Oil
PBS Video
1320 Braddock Place
Alexandria, VA 22314
Phone: (800) 344-3337
Color, 58 minutes, sound, 1/2" VHS, 1987.

Oil history of winners and losers during the time of John D. Rockefeller, and how he created the Standard Oil Company.

Oil: Episode 4—Rise of OPEC
PBS Video
1320 Braddock Place
Alexandria, VA 22314
Phone: (800) 344-3337
Color, 60 minutes, sound, 3/4" U-matic, 1/2" VHS, 1987.

Discusses the development of the OPEC producers cartel to counterbalance the economic power of the major oil corporations. This film charts the rise of OPEC.

Oil: Episode 8—Global Gamble
PBS Video
1320 Braddock Place
Alexandria, VA 22314
Phone: (800) 344-3337
Color, 58 minutes, sound, 1/2" VHS, 1987.

Presents the oil glut of the 1980s in Norway, Africa, China, and other countries. The question is, are the reserves sufficient to ensure oil supplies in the future?

Oil: Finds for the Future
Media Guild
11722 Sorrento Valley Road, Suite E
San Diego, CA 92121
Phone: (619) 755-9191
Color, 25 minutes, sound, 1/2" VHS, 1985.

Discusses the development of digital seismic recording as it influences oil exploration and discovery. Examines efforts to uncover resources and create technology to make these discoveries economically viable.

The Oil Age
Journal Films
930 Pitner Avenue
Evanston, IL 60202
Phone: (312) 328-6700
Color, 26 minutes, sound, Beta, 1/2" VHS, 1984.

Documents the long history of oil and how the industry became a major component of the industrialized twentieth century. Discusses the economic and political power and the influences that oil producers have on world markets. Looks at the economic depression of the late 1970s and early 1980s and its relation to oil prices.

Oil and Gas Production
University of Texas
Petroleum Extension Service
Balcones Research Center
10100 Burnet Road
Austin, TX 78758
Phone: (512) 471-3154
Color, 24 minutes, sound, 3/4" U-matic, 1/2" VHS, 2 × 2 slide, 1985.

This film shows oil and gas reservoirs and gives steps for producing and processing oil, gas, and water from the reservoir to the pipeline.

Oil and Sudden Wealth (Gulf Countries)
Coronet/MTI Film and Video
Supplementary Educational Groups
Simon and Schuster Communications
108 Wilmot Road
Deerfield, IL 60015
Phone: (800) 323-5343
Color, 28 minutes, sound, 16mm, 1978.

Investigates the lives of the people of the Gulf Countries (Middle East) and how they have been changed by great oil wealth. Considers long-term thinking of national policymakers.

Oil, Arms and the Gulf
Center for Defense Information
1500 Massachusetts Avenue, NW
Washington, DC 20005
Phone: (202) 862-0700
Fax: (202) 862-0708
Color, 28 minutes, sound, 1/2" VHS, 3/4" U-matic, 1991.

Develops the long-term implications of U.S. and European military involvement in Saudi Arabia and the Persian Gulf in response to the

Iraqi invasion of Kuwait. This is a balanced presentation with fine historical aspects. Many interviews are presented that are intelligent and thoughtful.

The Oil Kingdoms, Part I—Kings and Pirates
University of Utah
Educational Media Center
207 Milton Bennion Hall
Salt Lake City, UT 84112
Color, 57 minutes, sound, 1/2" VHS, 1983.

Presents a historical look at the discovery of oil in the Middle East and the incredible amount of petrodollars that have moved the Persian Gulf countries into the modern world.

The Oil Kingdoms, Part 2—Petrodollar Coast
University of Utah
Educational Media Center
207 Milton Bennion Hall
Salt Lake City, UT 84112
Color, 57 minutes, sound, 1/2" VHS, 1983.

Presents a look at the countries, the cities, and the societies that have emerged in the Persian Gulf and the rapid influx of wealth into the accompanying modernization.

The Oil Kingdoms, Part 3—Sea of Conflict
University of Utah
Educational Median Center
207 Milton Bennion Hall
Salt Lake City, UT 84112
Color, 58 minutes, sound, 1/2" VHS, 1983.

A study of the smaller Arab countries and their political and economic position in the world today.

Oil 105
WNET-TV
356 W. 58th Street
New York, NY 10019
Color, 30 minutes, sound, 3/4" U-matic, 1/2" VHS, video, 1985.

Attempts to demystify the money aspect so that both small and large businesses and the public understand new social and economic trends. Reports on economic trends and discoveries of the day.

Petroleum—River of Energy
Access Network
295 Midpark Way, SE
Calgary, Alberta
Canada T2X 2AB
Phone: (403) 256-1100
Fax: (403) 256-6837
Color, 59 minutes, sound, 3/4" U-matic, 1/2" VHS, 1988.

Describes, in simple language, the stages of exploration, acquisition of rights, drilling, production and marketing, natural gas processing, oil refining, petrochemicals, and transportation. Investigates specialized recovery techniques used for heavy oil, tar sands, and offshore drilling. Reviews the development of the Canadian oil industry.

Petroleum Geology
International Training Company
3301 Allen Parkway
P.O. Box 3881
Houston, TX 77001
Phone: (713) 529-5928
Color, 43 minutes, sound, Beta, 1/2" VHS, 3/4" U-matic, 1984.

A study of petroleum geology. There are four programs: (1) Eastern United States, (2) Western United States, (3) World and, (4) World explorations.

Petroleum's Progress
National Film Board of Canada
1251 Avenue of the Americas
New York, NY 10020-1173
Phone: (212) 586-5131
Color, 6 minutes, sound, Beta, 1/2" VHS, 3/4" U-matic, 1981.

An entertaining history of oil, oil by-products, and the creation of the petrochemical industry. The oil dependency question is raised.

Coal

Coal
University of Illinois
Visual Aids Service
1325 South Oak Road
Champaign, IL 61820
Color, 20 minutes, sound, 3/4" U-matic, 1/2" VHS, 1/2" Beta, 1983.

Describes coal gasification. Examines coal as a recoverable fuel source and describes the process characterizing its stages and properties.

Coal: Bridge to the Future

Exxon Corporation
1251 Avenue of the Americas
New York, NY 10020
Color, 29 minutes, sound, 16mm, 1/2" VHS, 1980.

It traces the history of coal through animation. Mining, grading, transportation of coal, and the effects of coal on the ecology are examined.

Coal: Solution or Pollution?

CRM Films
2233 Faraday Avenue
Carlsbad, CA 92008
Phone: (619) 431-4800
Color, 30 minutes, sound, 3/4" U-matic, 1/2" VHS, 1980.

This film stresses the technical problems of coal preparation, such as removing sulfur to clean the coal for environmental protection.

Coal—The Sleeping Giant

University of Colorado
Educational Media Center
Boulder, CO 80309
Color, 29 minutes, sound, 16mm, 1979.

With the abundance of coal in the United States, the problems faced are: How can it be mined with little impact on the land and health of miners? How can it be cleaned so it will not produce pollutants? How can the rising costs of equipment, manpower, and delivery be met? How can coal be turned into clean liquid products?

A Coal Operator's Turn

Pennsylvania State University
Audio-Visual Services, Media and Learning
Resources Division
University Park, PA 16802
Color, 29 minutes, sound, 3/4" U-matic, 1981.

Examines the Bradford Coal Company, a family owned and operated plant. Alan Walker, president of the company, expresses his opposition to government regulations that make it hard to operate his business economically and efficiently. It also shows how he works with Pennsylvania's Department of Environmental Resources on issues such as permits, pollution, and complaints.

Coal Strip Mine and Reclamation
Kent State University
Audio Visual Services
330 Library
Kent, OH 44242
Color, 30 minutes, sound, 3/4" U-matic, 1/2" VHS, 1982.

This film, produced at the Simco-Peabody mine, Coshocton County, Ohio, tells the story of coal strip mining including equipment, laws in Ohio, and reclamation.

Coal to Kilowatts
AIMS Media
6901 Woodley Avenue
Van Nuys, CA 91406
Phone: (818) 785-4111
       (800) 367-2467
Color, 9 minutes, sound, 16mm, 1979.

This film shows how ecological problems caused by strip mining can be solved. It examines open pit strip mining as a source of energy.

Energy Resources—Coal
Media Guild
11722 Sorrento Valley Road, Suite E
San Diego, CA 92121
Phone: (619) 755-9191
Color, 25 minutes, sound, 1/2" VHS, 1984.

A comprehensive examination of coal-field geology and modern techniques of coal mining. Examines different types of coal seams and illustrates the advantages and limitations of mechanized mining.

Natural Gas

Anti-Trust and You: A Guide for the Natural Gas Industry
American Gas Association
1515 Wilson Boulevard
Arlington, VA 22209
Phone: (703) 841-8461
Fax: (703) 841-8406
Color, 24 minutes, sound, 1/2" VHS, 3/4 U-matic, 1989.

Provides information of what is and what is not an antitrust situation using vignettes of real issues and case studies.

Natural Gas
L & K International Video Training
295 Evans Avenue
Toronto, ON M8Z 5P9
Canada
Phone: (416) 252-6407
Color, 60 minutes, sound, Beta, $^1/_2''$ VHS, $^3/_4''$ U-matic, 1988.

Describes the operation of a natural gas utility including pressure control, gas measurement, and safety in the natural gas industry.

Hydroelectric

Energy Conversion
Phoenix/BFA Films and Video, Inc.
470 Park Avenue South
New York, NY 10016
Color, 12 minutes, sound, 16mm, $^1/_2''$ VHS, 1985.

This shows energy conversion at a hydropower station where mechanical energy is converted into electrical energy.

Water and Energy
University of Illinois
Visual Arts Service
1325 South Oak
Champaign, IL 61820
Color, 17 minutes, sound, $^1/_2''$ VHS, $^3/_4''$ U-matic, Beta, 1980.

Describes some of the unusual properties of water and discusses them in terms of hydrogen bonding. The heating curve of water is shown and consequent energy changes of water.

Nuclear Energy

The Atom: A Closer Look
Disney Educational Productions
500 S. Buena Vista Street
Burbank, CA 91521
Phone: (800) 621-2131
Color, 30 minutes, sound, Beta, $^1/_2''$ VHS, EJ, $^3/_4''$ U-matic, 1990.

Nuclear energy and how it will be used in the future is shown. The structure of the atom is examined.

Bound by the Wind
Video Project
5332 College Avenue, Suite 101
Oakland, CA 94618
Phone: (510) 655-9050
Color, 40 minutes, sound, Beta, 1/2" VHS, 3/4" U-matic, 1991.

Tells the story of a group of people who live near nuclear test sites and are trying to halt international nuclear testing.

Chernobyl: Chronicle of Difficult Weeks
Video Project
5332 College Avenue, Suite 101
Oakland, CA 94618
Phone: (510) 655-9050
Color, 54 minutes, sound, Beta, 1/2" VHS, 3/4" U-matic, 1986.

A chronicle of the three months following the Chernobyl disaster. Some of the film is poor quality due to radiation at the site.

Deapsmith, A Nuclear Folktale
Video Project
5332 College Avenue, Suite 101
Oakland, CA 94618
Phone: (510) 655-9050
Color, 43 minutes, sound, Beta, 1/2" VHS, 3/4" U-matic, 1990.

The story of Deapsmith County.

A Dream of Peace
Video Project
5332 College Avenue, Suite 101
Oakland, CA 94618
Phone: (510) 655-9050
Color, 12 minutes, sound, Beta, 1/2" VHS, 3/4" U-matic, 1989.

The story of New Zealand's decision to declare itself a nuclear-free zone.

Energy: The Nuclear Alternative (Second Edition)
Churchill Films
12210 Nebraska Avenue
Los Angeles, CA 90025
Phone: (213) 207-6600
          (200) 334-7830

Fax: (213) 207-1330
Color, 22 minutes, sound, 16mm, 1/2" VHS, 1980.

Traces the history of the nuclear power program from the original government plans to the curtailment of activity in 1980. Describes how a reactor powers a turbogenerator. Discusses plant safety and waste management. Examines the failure at Three Mile Island and presents views of government officials, scientists, engineers, industry, and concerned citizens.

Energy from the Crust
Media Guild
11722 Sorrento Valley Road, Suite E
San Diego, CA 92121
Phone: (619) 755-9191
Color, 24 minutes, sound, 1/2" VHS, 1985.

Animated graphics show the hydrothermal activity that carried uranium up from the earth's center and formed ore deposits near the surface. It shows how deposits are located and how they will be extracted.

The Energy Problem—The Nuclear Solution
Encyclopaedia Britannica Educational Corp.
310 S. Michigan Avenue
Chicago, IL 60604
Phone: (312) 347-7900
       (800) 621-3900
Color, 16 minutes, sound, 1/2" VHS, 1986.

Presents a survey of the worldwide problem of dwindling energy supplies and shows how nuclear fission and nuclear fusion can serve as energy solutions.

Nuclear Energy: A Perspective
Modern Talking Picture Service
5000 Park Street, N
St. Petersburg, FL 33709
Color, 28 minutes, sound, 3/4" U-matic, 1984.

Tells the story of uranium from the search for ore to fueling of nuclear reactors. Also looks at the future of breeder reactors.

The Nuclear Fuel Cycle
Encyclopaedia Britannica Corp.
310 S. Michigan Avenue
Chicago, IL 60604
Phone: (312) 347-7900
       (800) 621-3900
Color, 21 minutes, sound, 1/2" VHS, 1986.

Describes the fuel cycle beginning with the mining of uranium and follows the stages through enrichment, fabrication use in the thermal reactor, reprocessing, and the disposal of the waste.

Nuclear Fusion—Energy for the 21st Century
Encyclopaedia Britannica Corp.
310 S. Michigan Avenue
Chicago, IL 60604
Phone: (312) 347-7900
        (800) 621-3900
Color, 25 minutes, sound, 1/2" VHS, 1983.

Discusses present-day nuclear fusion, the hope that nuclear fusion will evolve for the future, and the uniting of deterium and tritinum nuclei— a process that releases enormous amounts of energy.

Nuclear Spin-Off
Encyclopaedia Britannica Corp.
310 S. Michigan Ave.
Chicago, IL 60604
Phone: (312) 347-7900
        (800) 621-3900
Color, 28 minutes, sound, 1/2" VHS, 1986.

Explains the positive side of the nuclear paradox, showing how nuclear research offers benefits in such fields as industry, medicine, and transportation.

Nuclear Theory and Energy
Britannica Films
310 S. Michigan Avenue
Chicago, IL 60604
Phone: (312) 347-7958
Fax: (312) 347-7966
Color, 24 minutes, sound, Beta, 1/2" VHS, 3/4" U-matic, 1988.

The viewer is taken outside an atomic nucleus to witness what is actually happening during such events as fission and radioactive particle releasing.

Nuclear Waste Isolation: A Progress Report
U.S. Department of Energy
Office of Public Affairs
1000 Independence Avenue, SW
Washington, DC
Color, 26 minutes, sound, 16mm, 1981.

Examines the U.S. Department of Energy's attempts to find a safe disposal system for highly radioactive wastes, with a focus on the concept of placing the wastes in deep underground repositories. Describes temporary storage methods.

Renewable Energy

Energy
Journal Films
930 Pitner Avenue
Evanston, IL 60202
Phone: (312) 328-6700
Color, 27 minutes, sound, 16mm, 1/2" VHS, 1985.

Discusses alternative energy sources such as geothermal stations, wind generators, solar and chemical energy, tidal power, and power from controlled nuclear fusion and the steps taken to harness the potential power.

Energy: The Alternatives
Media Guild
11722 Sorrento Valley Road, Suite E
San Diego, CA 92121
Phone: (619) 755-9191
Color, 25 minutes, sound, 1/2" VHS, 16mm, 1984.

Considers energy alternatives to fossil fuels, including harnessing the ocean's tidal cycle, utilizing geothermal resources, harvesting the wind, and developing methanol fuel to replace gasoline.

The Energy Balance
Coronet/MTI Film and Video
Supplementary Education Group
Simon and Schuster Communications
108 Wilmot Road
Deerfield, IL 60015
Phone: (800) 323-5343
Color, 15 minutes, sound, 16mm, 3/4" U-matic, 1/2" VHS, 1986.

Discusses the balance between incoming solar energy and energy returned to space. Changes in ice and snow, volcanic eruptions, and effects of human activities on the greenhouse effect are shown.

Energy from the Sun (Second Edition)
Encyclopaedia Britannica Educational Corp.
310 S. Michigan Avenue
Chicago, IL 60604

Phone: (312) 347-7900
      (800) 621-3900
Color, 17 minutes, sound, 16mm, 1980.

The film presents a survey of the various ways modern technology is used to capture and store the sun's energy as an alternative to fossil fuels and as the key to developing energy sources for the future.

Energy: New Source (Second Edition)
Churchill Films
2210 Nebraska Avenue
Los Angeles, CA 90025
Phone: (213) 207-6600
      (200) 334-7830
Fax: (213) 207-1330
Color, 25 minutes, sound, 16mm, 1/2" VHS, 1980.

Surveys new sources of energy that have the greatest potential: solar, geothermal, nuclear fusion, and synthetic fuels. Includes a survey of water and space heating, biomass sources, the oceans, wind power, and photovoltaic conversion.

Energy Seekers
Centron Educational Films
Supplementary Educational Groups
Simon and Schuster Communications
108 Wilmot Road
Deerfield, IL 60015
Phone: (800) 323-5343
Color, 11 minutes, sound, 16mm, 1980.

Surveys the present and future promise of new energy sources. It shows a windmill as a source of energy and a solar thermal power system producing electricity.

Solar

Solar Comfort
California Energy Commission
Sacramento, CA 95815
Color, 23 minutes, sound, 16mm optical sound, 1980.

This film shows how proper home design can provide solar heating and cooling with a minimum of energy used. It includes interviews with owners of passive solar homes.

Solar Energy
Barr Films
12801 Schabarum Avenue
P.O. Box 7878
Irwindale, CA 91107
Phone: (818) 338-7878
Color, 23 minutes, sound, 3/4" U-matic, 1/2" VHS, 1990.

The film shows how sun produces heat and light and discusses the ways in which plants and animals use solar energy. Looks at uses for solar power.

Solar Energy: How It Works
Churchill Films
12210 Nebraska Avenue
Los Angeles, CA 90025
Phone: (213) 207-6600
       (800) 334-7830
Fax: (213) 207-1330
Color, 16 minutes, sound, 16mm, 1980.

This piece shows how important it is to develop solar energy to produce heat, light, and power. It shows how the sun's energy can be used by concentration, absorption, and by conversion to electricity.

Solar Energy Now
PBS Video
1320 Braddock Place
Alexandria, VA 22314-1698
Phone: (800) 344-3337
Fax: (703) 739-0775
Color, 30 minutes, sound, 3/4" U-matic, 1/2" VHS, 1982.

Looks at how solar energy systems work in the home. It was taped at the Farallones Institute's Integral Urban House in Berkeley, California, and shows the installation of home solar energy heating units.

The Solar Film
Pyramid Films and Video
2801 Colorado Avenue
Santa Monica, CA 90404
Phone: (213) 828-7577
Color, 9 minutes, sound, 16mm, 1/2" VHS, 1980.

This animated film depicts the relationships between the sun and mankind. It shows our use of sun created fossil fuels. The film was produced by Robert Redford to explain what solar energy is and how it is a good source of future power.

The Solar Frontier
Bullfrog Films, Inc.
Oley, PA 19547
Phone: (215) 779-8226
         (800) 543-3764
Color, 24 minutes, sound, 16mm, 1977.

This film features three solar houses, all with different systems to provide comfort at less expense than conventional fuels. Cost, technology, and performance are explained in the film.

The Solar Horizon
AIMS Media
6901 Woodley Avenue
Van Nuys, CA 91406
Phone: (818) 785-4111
         (800) 367-2467
Color, 10 minutes, sound, 16mm, 1981.

In ten minutes this film shows how to harness energy from the sun by use of active and passive collectors. Also shown is the production of electricity from the sun's rays by the use of photovoltaic cells.

The Solar House
National Film Board of Canada
1251 Avenue of the Americas
New York, NY 10020
Phone: (212) 586-5131
Color, 12 minutes, sound, 16mm, 1/2" VHS, 1985.

Explains the design and operation of solar heating systems. Describes the difference between active and passive heating. Explains the economics of using solar energy for heating.

Solar Power
Handel Film Corporation
8730 Sunset Boulevard
West Hollywood, CA 90069
Color, 20 minutes, sound, 16mm, 3/4" U-matic, 1/2" VHS, 1980.

Shows solar power used for passive heating, thermal heating, and photovoltaic production of electricity.

Index